Palgrave Studies in Impact Finance

Series Editor
Mario La Torre, Department of Management, Sapienza University
of Rome, Rome, Italy

The *Palgrave Studies in Impact Finance* series provides a valuable scientific 'hub' for researchers, professionals and policy makers involved in Impact finance and related topics. It includes studies in the social, political, environmental and ethical impact of finance, exploring all aspects of impact finance and socially responsible investment, including policy issues, financial instruments, markets and clients, standards, regulations and financial management, with a particular focus on impact investments and microfinance.

Titles feature the most recent empirical analysis with a theoretical approach, including up to date and innovative studies that cover issues which impact finance and society globally.

More information about this series at
http://www.palgrave.com/gp/series/14621

Maram Ahmed

Innovative Humanitarian Financing

Case Studies of Funding Models

Maram Ahmed
Faculty of Law & Social Science
School of Oriental and African Studies
London, UK

ISSN 2662-5105   ISSN 2662-5113   (electronic)
Palgrave Studies in Impact Finance
ISBN 978-3-030-83208-7   ISBN 978-3-030-83209-4   (eBook)
https://doi.org/10.1007/978-3-030-83209-4

Cover illustration: Volodymyr Burdiak/Alamy Stock Photo

This Palgrave Macmillan imprint is published by the registered company Springer Nature Switzerland AG
The registered company address is: Gewerbestrasse 11, 6330 Cham, Switzerland

Dedicated to my wonderful family.

CONTENTS

CHAPTER 1

Introduction

In 2016, there were 97 million people across 40 countries who were in need of humanitarian assistance, according to the United Nations Office for the Coordination of Humanitarian Assistance (OCHA) (Hoffman 2016). Overall, just five crises (in Syria, Yemen, South Sudan, Iraq and Ethiopia) accounted for more than 53% of all funding allocated to specific emergencies in 2016 (Spiegel et al. 2018). Furthermore, an estimated 65 million people were displaced (or one in every 113 people worldwide), many of which are in "protracted refugee situations" where they have been displaced for over five years and with some situations lasting for over thirty years (Hoffman 2016). By the end of 2016, international humanitarian assistance, including assistance related to conflict, natural disasters, internally displaced persons and non-displaced persons, reached a record high of US$27.3 billion (Spiegel et al. 2018). Overall, a total of five high income government donors (with the exception of Turkey) contributed to 65% of the total humanitarian funding in 2016, with European countries having provided 53% combined and the United States provided 31% (Spiegel et al. 2018).

Against this backdrop, world leaders convened at the first ever World Humanitarian Summit that took place in Istanbul in 2016, to discuss pressing concerns and much-needed reforms needed for the international humanitarian architecture. The summit put the humanitarian funding gap

© The Author(s), under exclusive license to Springer Nature Switzerland AG 2021
M. Ahmed, *Innovative Humanitarian Financing*,
Palgrave Studies in Impact Finance,
https://doi.org/10.1007/978-3-030-83209-4_1

at the top of the agenda with the release of a report to the UN Secretary General from the High-Level Panel on Humanitarian Financing. Of the reforms announced at the World Humanitarian Summit, the "Grand Bargain" initiative was created with a commitment from major donors and humanitarian organizations to work toward a series of goals—many of which centered on the humanitarian funding relationship—with the aim of improving efficiency (Stoddard et al. 2017). These reforms included the reduction of duplication and management costs, the reduction of donor earmarking, the simplification of reporting requirements, the improvement of financial transparency, the increase in direct funding to national first responders, the increase in programs that are cash-based and the increase of multi-year funding (Stoddard et al. 2017). The landmark Summit marked a shift toward more deliberate efforts to reduce needs, a commitment to political will and leadership in attempting to prevent and end conflict, in addition to a pledge to bridge the divide across humanitarian, development, human rights, peace and security interventions efforts (Thulstrup et al. 2020).

Although in recent times the ongoing COVID-19 pandemic constitutes as a global crisis with far-reaching consequences, prior to the pandemic, there was no shortage of people in need of humanitarian assistance. Humanitarian organizations and governments in less developed countries have long dealt with a series of shocks and crises that required substantial funding in order to be addressed. From protracted crises, forced displacement, natural disasters, conflicts, extreme poverty, food and water insecurity—there are an increasing number of people in need of humanitarian assistance and regardless of the pandemic, many humanitarian crises remain significantly underfunded. However, in dealing with the effects of the COVID-19 pandemic, governments are prioritizing the needs of their citizens and national interests. With the pandemic forecasted to negatively impact the global economy for several years to come, there will inevitably be cuts to humanitarian aid budgets and less funding available for humanitarian purposes.

Prior to the pandemic, an estimated 79.5 million were forcibly displaced around the world (UNHCR 2020) with the number expected to increase. Conflicts are also no longer temporary as in most cases, they are protracted, complex and inflict prolonged long-term suffering. And with the increasing and ongoing impacts of climate change, humanitarian crises will be compounded and more complex. For instance, in 2019 alone there were 396 natural disasters which resulted in 11,755 deaths and impacted 94.9 million people (CRED 2019), moreover, as a result of crop failure, water scarcity and sea level rise, it is estimated that 143 million people around the world are expected to be forcibly displaced (Rigaud et al. 2018). Also, in conflicts, millions of children have been killed, injured or psychologically harmed and according to the United Nations Children's Fund (UNICEF), the number of countries where conflicts are impacting children is the highest it has been since the adoption of the Convention on the Rights of the Child in 1989 (UNICEF 2019; Jaff 2020). Moreover, in 2019, UNICEF released an emergency appeal for USD$4.2 billion—a number that is more than triple the amount the organization requested in 2010 (UNICEF 2019).

As humanitarian needs increase, so do financial demands; therefore, traditional humanitarian funding is struggling to keep up with the explosion of needs and cannot solely be relied upon as humanitarian organizations are no longer able to rely on traditional donor assistance alone to meet increasing funding needs. Given this, more humanitarian organizations are evolving their funding methods and are expanding their funding pools by tapping private financial sources and crowding in non-governmental actors in a variety of ways, in order to mobilize and channel greater funding to crises. As such, more humanitarian organizations are increasingly collaborating with the private sector and utilizing innovative financial products and services. This is as with the widening deficit and growing humanitarian needs, it is evident that no single actor can plug the deficit.

Several years on from the World Humanitarian Summit, there is a renewed urgency and need for effective humanitarian funding to be provided for millions in need across the world. However, there is a shortfall of adequate funding available leaving many humanitarian organizations and governments around the world, often struggling to deal with, and contain humanitarian crises. In the past couple of years, as the humanitarian landscape (funding needs in particular) rapidly changes, there has been an increased interest in exploring and utilizing innovative funding mechanisms. Innovative humanitarian financing can be defined as the channeling, mobilization and provision of non-traditional donor funding for humanitarian purposes and needs.

As alternative funding models are required from innovative financing sources to address ever-increasing humanitarian needs, this presents a number of challenges and opportunities. As more humanitarian organizations test, pilot and search for sustainable and innovative sources of funding, this book will offer new perspectives on the emergent innovative humanitarian financing field. Furthermore, the lessons learnt from these innovative humanitarian financing mechanisms that aim to develop and expand the funding pools of humanitarian organizations, will be scrutinized.

AIMS AND OBJECTIVES

Due to the widening deficit in humanitarian funding, and a growing recognition of the potential of innovative funding mechanisms, there has been a surge of interest in innovative humanitarian financing in recent years. Although it is argued that the mobilization of innovative humanitarian financing tools can help close the funding gap and provide much-needed financing for humanitarian crises, there is limited research and understanding of how these funding tools work, the opportunities they offer, their challenges in addition to how they fit into the wider humanitarian funding landscape.

This book is among the first to assess a set of innovative financing mechanisms that have been transforming the humanitarian sector in recent years. Furthermore, this book explores the key opportunities, challenges and future potential of this emergent field. This book attempts to unveil the changes that are happening in the humanitarian sector in regard to innovative finance, in a context which is both historical and forward-looking. The aims of this book are threefold: (1) outline the main features of humanitarian funding and explore a range of non-traditional financing mechanisms and approaches being used to tap new funding pools; (2) analyze landmark innovative humanitarian financing models that so far have not been collectively explored in the policy, industry and academic sphere; (3) provide a road map of priorities and policy strategies for further development.

RESEARCH DESIGN, METHODOLOGY AND CONCEPTUAL FRAMEWORKS

The emergent field of innovative humanitarian financing is at the early stages of being understood, researched and analyzed; therefore, given the lack of data available and the nature of the topic, this book will apply a case study approach as the general research methodology. Moreover, the analytical framework used in this research will be a qualitative approach whereby an exploratory analysis will be conducted on a selected number of cases. The objective of using such an approach is to provide a holistic understanding of the issues and opportunities that arise when structuring and implementing financial mechanisms for a humanitarian context.

As a research methodology, a case study "investigates a contemporary phenomenon within its real-life context; when the boundaries between phenomenon and context are not clearly evident; and in which multiple sources of evidence are used" (Yin 1984), moreover, this methodology enables the building and developing of theories as well as enables us to draw inferences, therefore it is better suited for the overall aim of this book. The cases used in this study were selected as they either presented interesting funding mechanisms or were the first of their kind.

STRUCTURE OF THE BOOK

To give some context, the next chapter of the book will give a background by analyzing how the traditional humanitarian funding system currently operates. Also, this chapter will discuss the key challenges facing the traditional humanitarian financing system and how it paved the way for the rise of innovative humanitarian finance.

The next part of the book will narrow the scope of the study by focusing on and evaluating, a select number of innovative financing mechanisms that have been used in the humanitarian sector in recent years. Through case studies, this section will examine non-traditional funding mechanisms and scrutinize their role and effectiveness as well as assess their potential, opportunities and challenges.

Chapter 3 will assess and evaluate the role of innovative vaccine and immunization funding. In particular, this chapter will focus on advance market commitments pioneered by GAVI.

Chapter 4 will assess the role of performance-based financing in the humanitarian sector and more specifically, impact bonds. In order to assess the viability of this funding mechanism, this chapter will evaluate the world's first Humanitarian Impact Bond launched by the International Committee of the Red Cross (ICRC).

Chapter 5 will assess the increasing role of Islamic financial mechanisms, in particular, Islamic social financial tools such as Zakat, in humanitarian response through conducting a case study on the International Federation of the Red Cross and Red Crescent's (IFRC) drought-assistance program that utilized this mode of funding.

With the rise of climate-induced humanitarian crises, Chapter 6 will investigate the role of financial inclusion in addressing immediate vulnerability as well as enhancing resilience in humanitarian contexts, and in particular Islamic financial inclusion. Moreover, this chapter will present a case study of the first private cooperative to offer Islamic financial products and services in Wajir, Kenya as an attempt to enhance financial inclusion, that was implemented by humanitarian organizations MercyCorps.

Chapter 7 will explore the use of risk financing instruments and in particular, disaster risk insurance, and its potential as a funding tool to provide financial security against disasters moreover, the chapter will scrutinize the African Risk Capacity—a sovereign disaster risk insurance mechanism.

The last part of the book will discuss some of the future trends in humanitarian financing.

Chapter 8 will review how digital funding tools are transforming the humanitarian sector. This chapter will explore the potential and opportunities of using digital financial tools for humanitarian purposes.

Chapter 9 will conclude the book's overall arguments, findings and contributions. It will also provide further research suggestions and policy priorities needed to further develop innovative humanitarian financing.

REFERENCES

Centre for Research on the Epidemiology of Disasters (CRED). (2019). Natural disasters 2019. Brussels: Centre for Research on the Epidemiology of Disasters. Retrieved from https://reliefweb.int/sites/reliefweb.int/files/resources/ND19.pdf.

Hoffman, P. J. (2016). Humanitarianism in treatment: Analyzing The World Humanitarian Summit. FUNDS Briefing, 41.

Jaff, D. (2020). Financing and resolving the ever-increasing humanitarian crises. *Medicine, Conflict and Survival*, 36(2), 129–131.

Rigaud, K. K., de Sherbinin, A., Jones, B., Bergmann, J., Clement, V., Ober, K., Schewe, J., et al. (2018). *Groundswell: Preparing for internal climate migration*. Washington, DC: World Bank. Retrieved from: https://openknowledge.worldbank.org/bitstream/handle/10986/29461/WBG_ClimateChange_Final.pdf.

Spiegel, P., Chanis, R., and Trujillo, A. (2018). Innovative health financing for refugees. *BMC Medicine*, 16(1), 1–10.

Stoddard, A., Poole, L., Taylor, G., and Willitts-King, B. (2017). Efficiency and inefficiency in humanitarian financing. Retrieved from: https://www.humani tarianoutcomes.org/sites/default/files/publications/humanitarian_financing_ efficiency_.pdf.

Thulstrup, A. W., Habimana, D., Joshi, I., and Oduori, S. M. (2020). Uncovering the challenges of domestic energy access in the context of weather and climate extremes in Somalia. *Weather and Climate Extremes*, 27, 100185.

United Nations Children's Fund (UNICEF). (2019). UNICEF Issues Largest-ever Emergency Funding Appeal to Reach Historic Numbers of Children in Need. Retrieved from: https://www.unicef.org.uk/press-releases/unicef-iss ues-largest-ever-emergency-funding-appeal-to-reach-historic-numbers-of-chi ldren-in-need/.

UNHCR. (2020). Figures at a glance. Retrieved from: https://www.unhcr.org/ figures-at-a-glance.html.

Yin, R. K. (1984). *Case study research: Design and methods*. London: Sage.

CHAPTER 2

Humanitarian Funding in Crisis and the Rise of Innovative Humanitarian Finance

Humanitarian aid is defined as "the impartial, independent and neutral provision of aid to those in immediate danger" (Rysaback-Smith 2015). If a major disaster strikes a country and the local emergency services are overwhelmed and struggling to deal with the disaster, external help is usually requested whereby humanitarian organizations respond to the crisis by delivering aid such as food, medical supplies and shelter to those in need (Vitoriano et al. 2011). In most cases, national and international aid agencies tend to provide technical and financial assistance to victims (Safarpour 2018) and humanitarian funding is typically provided by a donor to a humanitarian organization for direct implementation or it is outsourced to implementing partners (Spiegel et al. 2018).

When a country is struck by a large disaster, international humanitarian organizations tend to respond immediately to implement disaster response programs (Turrini et al. 2020). These programs are usually multimillion-dollar programs that tend to cover a variety of activities such as procurement, asset management and logistics (van Wassenhove 2006). Humanitarian program are not implemented in isolation; instead, they operate in a social arena where context, policy and various actors are all interconnected (Bakewell 2000; Walt 1994). The collaboration and involvement of the community in all phases in disaster management is essential in order to successfully achieve disaster management (Safarpour

© The Author(s), under exclusive license to Springer Nature   9
Switzerland AG 2021
M. Ahmed, *Innovative Humanitarian Financing*,
Palgrave Studies in Impact Finance,
https://doi.org/10.1007/978-3-030-83209-4_2

et al. 2020); however, the type of assistance provided can differ depending on the characteristics of each country (Jahangiri et al. 2011).

The aim of this chapter to evaluate the reasons as to why humanitarian funding is in crises and why innovative humanitarian financing is on the rise. The first part of the chapter will discuss some of the drivers that are widening the humanitarian financing gap such as conflict, forced displacement and environmental and humanitarian issues. The second part of this chapter will discuss some of the issues in humanitarian financing and recent trends.

WIDENING GAP BETWEEN HUMANITARIAN NEEDS AND AVAILABLE FUNDING

One of the most significant challenges international humanitarian organizations are facing is the widening gap between the demand for humanitarian aid and the funding they receive (Jahre and Heigh 2008; Wakolbinger and Toyasaki 2011), and the gap is increasingly growing (Stumpf et al. 2017). Every year as the number of humanitarian crises increase, the gap between humanitarian needs and humanitarian funding further widens. According to Scott (2015), since 2000, requests for humanitarian support have increased by 660%, from US$2.9 billion (adjusted for inflation) to US$18.6 billion in 2015.

Some of the drivers that result in increases in money being spent on humanitarian aid are "environmental, social and demographic, and geopolitical shifts, coupled with the increasing influence of technology, a globalized economy, and rising inequality" (Scott 2014). Furthermore, with a range of factors such as escalating violence within and across states, climate change and other environmental disasters—an increase in the number of victims and displaced persons is anticipated (Hoffman 2016). With challenges such as protracted crises, climate change and a growing number of displaced populations, the humanitarian sector is facing unprecedented challenges (Opdyke et al. 2021). Now more than ever, there is a greater need for humanitarian aid to better assist those that have been impacted by conflicts and disasters.

The frequency of conflicts and the number of conflict-related deaths from the period between 2006 and 2016 sharply increased from 33 and 19,601 in 2006, to 49 and 102,000 in 2016 (UCDP 2018; Allansson et al. 2017); moreover, violent conflicts and spikes of violence in particular are a key driver of forced displacement (Brück et al. 2018). Natural disasters and forced displacements are compounded by state fragility and weak institutions resulting in complex and prolonged humanitarian emergencies (Brück and d'Errico 2019). It is important to note that conflicts vary significantly from each other as even the same conflict can have varying impacts between people across time and space (Brück and d'Errico 2019). In addition, the physical and the psychological impact of exposure to violent conflicts early in life could result in strongly adverse and often irreversible short-term and long-term impacts, which could then be transmitted through generations (Alderman et al. 2006; Akresh et al. 2012; Bundervoet et al. 2009; Singhal 2019). The challenge of humanitarian aid "if it aspires to durable solutions, is to address the subnational nature and dynamics of conflict"(Lemaitre 2018).

FORCED DISPLACEMENT

In order to address the large movements of refugees and migrants, in 2016 the UN General Assembly convened the Global Migration Summit. The aim of the summit was to address and help shape the way rapidly rising numbers of vulnerable people (as a result of crises and climate-related disasters) were being supported (Thulstrup et al. 2020). According to UNHCR (2019), by the end of 2019, there were 79.5 million forcibly displaced people globally due to drivers of migration and displacement such as persecution, conflict, violence, human rights violations or other events that have disturbed public order. Moreover, of the forcibly displaced people, the majority are displaced within their own country.

Ensuring an adequate humanitarian response to the displacement crisis and meeting the basic needs of internally displaced people can be a significant financial burden on fragile countries and on overstretched humanitarian system (Cazabat and Yasukawa 2020). For example, in the case of Somalia, the cost of meeting the needs of the country's 2.6 million internally displaced people for a year of displacement was over USD$1 billion in 2018 (Cazabat and Yasukawa 2020) which equates to 21% of the country's GDP (World Bank 2020). Given that the amount of funds requested by OCHA to meet humanitarian needs across the world are often significantly underfunded and not met, it not only severely limits the resources available to assist internally displaced people (Cazabat and Yasukawa 2020) but it makes it increasingly difficult to address the underlying issues of displacement.

Environmental Issues

For the past several decades, the potential link between climate change, migration and conflict has been discussed (Burrows and Kinney 2016) as climate change and extreme weather events have been a growing concern (Dellmuth et al. 2021), especially in the humanitarian sector. In particular, there are concerns that higher occurrences of climate and weather-related disasters will increase the risks of population displacement, social unrest and conflict (Brzoska 2019; Koubi et al. 2018). Moreover, in the future, climate change is expected to result in changes in the frequency, intensity and duration of extreme events such as heat waves, heavy rain, coastal flooding and drought and associated wildfires (IPCC 2018).

Although climate change is a global issue where developed and developing countries alike will be impacted, the burden will not be the same (Dellmuth et al. 2021), for example, since the 1970s, over 95% of deaths from climate-induced natural disasters have occurred in developing countries (Handmer et al. 2012). Frequent climate and weather extremes is also one of the key drivers of the growing number of conflicts that are becoming increasingly protracted and complex (Thulstrup et al. 2020). Growing unresolved conflicts over natural resources and land due to competition over resource access and use coupled with excess exploitation of natural resources—risks worsening existing resource scarcity and humanitarian crises (Abdi et al. 2008; Dehérez 2009).

The potential linkages and interconnectedness between weather and climate-related disasters, conflict and migration (Scheffran et al. 2012; Brzoska and Fröhlich 2016; Abel et al. 2019) are often referred to as the disaster–migration–violent conflict nexus (Burrows and Kinney 2016). According to Brzoska (2019), to some extent, natural disasters, violent conflict and migration reinforce each other with "causal chains that can run in both directions"; however, there are some links in the nexus that are stronger than others (Scheffran et al. 2012; Ide and Scheffran 2014; Ide et al. 2016; Schilling et al. 2017).

Natural disasters and violent conflicts are major drivers of population displacement (Brzoska 2019). Furthermore, violent conflict and migration are factors that increase vulnerability to disasters, for example, when humanitarian assistance is withheld for security reasons or when migrants settle in environments that are risky (Harris et al. 2013; Siddiqi 2018). Moreover, climate change-induced impacts, such as droughts, can negatively impact both natural resource availability and food security and as a consequence could lead to migration, resource competition and conflict, therefore acting as a threat multiplier (Hamro-Drotz 2014). For instance, according to a study carried out by Thulstrup et al. (2020), competition over natural resources was found to be a key driver of climate-induced conflict in Somalia that was exacerbated by weather and climate extremes.

PROTRACTED CRISES

Protracted crises describe crises where countries are in states of continual and repeated crisis as a result of underlying vulnerabilities (Contreras et al. 2020). According to the Scott (2015), protracted crises are "complicated political and/or operational environments, often involving conflicts, which drag on for years, or decades. They are spaces where development actions are required, but instead they consume the majority of humanitarian funding." In sub-Saharan Africa (SSA), 65% of states have experienced conflict since obtaining independence, and since 2000, there have been an average of eight to ten conflicts on the continent in any given year (Straus 2012). These conflicts not only directly impact the loss of and disrupt livelihood assets, but they also result in forced migration, forcing populations to be refugees or be internally displaced (Opdyke et al. 2021). The levels of global population displacement as a result of conflict surpass records since measurements began (UNHCR 2018), by the end of 2018, 70.8 million people were forcibly displaced globally with numbers expected to continue to rise (UNHCR 2019). Furthermore, in many states that are in contexts of protracted armed conflict, the capacity to deliver public services is low (Hoffmann and Kirk 2013).

Although they tend to share similar characteristics—such as long durations, the existence of violent conflict, weak governance or public administration, unsustainable livelihood systems and the breakdown of local institutions—protracted crises are not all alike (Brück and d'Errico 2019). Humanitarian actors should therefore, according to Brück et al. (2018), focus on how aid relief contributes to longer-term development. The timeline of crises is often difficult to define. When a crisis takes place, there is typically an initial and immediate response to save lives and provide basic necessities which is referred to by the UNHCR as an "acute stage" (Burde et al. 2015). Moreover, the average life length of conflicts is estimated to be 12 years and natural disasters can disrupt countries for up to 8 years (UNESCO 2011; Wedge 2008), therefore with the life cycles of natural disasters or conflicts averaging between 8 to 12 years, the initial humanitarian response may last only for months or several years in many cases.

HUMANITARIAN AID ALLOCATION

Official bilateral aid is usually perceived to be slow, bureaucratic and politically driven (Lancaster 2007). Previous empirical research conducted found that the allocation patterns of official aid are not only determined by the needs and performance of recipient countries, in terms of their development policies, but they are also determined by the political and economic interests of the government of the donor country (Alesina and Dollar 2000; Faye and Niehaus 2012; Hoeffler and Outram 2011). This finding has also been found in studies that solely focus on humanitarian aid (Annen and Strickland 2017; Bommer et al. 2018; Fink and Redaelli 2011; Raschky and Schwindt 2012). According to Turrent and Oketch (2009), literature on humanitarian aid allocation can be divided into three areas: (1) aid effectiveness, (2) absorptive capacity and (3) predictability of aid flows.

Aid effectiveness places emphasis on the need for aid recipient countries to have strong policy and institutional environments in order for aid income to result in economic growth (Burnside and Dollar 2000; Collier and Dollar 2001, 2002). Some research conducted on the impact of aid flows has shown that aid increases growth and consequently contributes to a reduction in the levels of poverty (Hansen and Tarp 2000). Burnside and Dollar (2000) found that aid tends to work better in countries that have better policy regimes; however, Hansen and Tarp (2000) suggest that aid can work and be effective in countries that have an unfavorable policy environment. Similarly, McGillivray (2005) suggests that the link between aid effectiveness—as defined in terms of the impact of aid on growth—and policy might be as consistent or systematic across countries that are at the bottom end of the policy scale.

Absorptive capacity indicates that some countries that are limited by their institutional capacities have a lower saturation point (Heller and Gupta 2002; Clemens and Radelet 2003; Killick 2004). Turrent and Oketch (2009) suggest there are clear indications that similar to other investments, aid has diminishing returns. Studies have indicated that an aid saturation point could reach between 15 and 45% of GDP, beyond which the marginal benefits of any additional aid flows become negative (De Renzio 2005). This is as the concept of aid saturation has been interpreted to indicate that aid has limited absorptive capacity, with recipient governments being limited in the amounts of aid they can effectively use (Clemens and Radelet 2003). Furthermore, research conducted by Clemens and Radelet (2003) shows that certain types of "short-impact aid," which includes budget support and sector budget support, tend to have a stronger impact on growth than aid that is taken as a whole.

Predictability of aid flows suggests that the beneficial impacts of aid can indeed be offset by high volatility and unpredictability (Edwards and van Wijnberg 1989; Gemmell and McGillivray 1998; Bulíř and Hamann 2003; Levin and Dollar 2005). Levin and Dollar (2005) found discrepancy in aid volatility as aid to fragile states was sent in spurts, where a fragile state can receive substantial aid flows in one year and the next year, donors move on to another country. This sort of volatility can have a damaging impact on growth and poverty reduction (Turrent and Oketch 2009), and for fragile states, inconsistent allocation exacerbates the issue by making aid flows unpredictable (McGillivray 2005).

LACK OF MULTI-YEAR FUNDING

Humanitarian aid is provided by several funding sources; however, the largest amount of humanitarian aid is often derived from governments. The survival of the majority of humanitarian organizations therefore depends on international donor funding; however, typical funding cycles tend to be short-term rather than more long-term and multi-year. According to Fabre (2017), multi-year funding refers to the type funding given "over two or more years for humanitarian assistance, including funding for multilateral organizations, a national disaster management agencies, the Red Cross and Red Crescent Movement and local and international non-governmental organizations (NGOs)." The benefits of multi-year humanitarian funding include lower operational costs, flexibility for early response, predictability, local capacity building and enhancing resilience, coherence with development and recovery programs and more effective programming (Cabot-Venton 2013).

Financing that is both multi-year and flexible is considered to be "a critical enabler of nexus approaches" (Poole and Culbert 2019) and an important element of the "new way of working" (Al-Mahaidi 2020).

This is as short-term funding is often insufficient enough to help achieve long-term resilience or address long-term humanitarian issues and their underlying causes such as protracted crises, forced displacement and environmental issues. Multi-year funding approaches on the other hand can enable humanitarian, development and peacebuilding actors to work together in order to achieve durable solutions (FAO 2017) as well as create funding predictability compared to short-term funding that fluctuates. Furthermore, at the early stages of an emergency, the ability to access flexible multi-year funding sources can lead to more durable, long-term solutions.

At the 2016 World Humanitarian Summit, humanitarian donors made commitments to shift to multi-year humanitarian funding rather than annual funding (Fabre 2017). However, delivering on this commitment will be difficult as the majority of donors tend to commit to funding cycles of 12–18 months (Scott 2015). Although responding to long-term humanitarian needs with short-term funding cycles allows organizations to focus on meeting immediate humanitarian needs, it can prevent more long-term effects such as addressing the root causes of those humanitarian issues (Fabre 2017). To address the long-term nature of protracted crises and complex humanitarian crises, there is a growing consensus on the need to shift to longer-term funding models. Moreover, there is also a need to develop multi-year humanitarian programs (SESRIC 2017).

Rise in Earmarked Funding

After a disaster occurs, humanitarian organizations have the choice of whether to offer donors the option to earmark or not (Aflaki and Pedraza-Martinez 2016), and if they choose to earmark, donors can then track their donations and can have greater control on how their donations are being used (Nunnenkamp and Ohler 2012). A donation is considered as earmarked when donors contribute "for a specific trust/pooled fund, program or project, often with additional geographic, thematic or other proscriptions (rather than providing 'core' or "unrestricted" resources to multilaterally agreed budgets)" (Otto Baumann 2020).

There are several reasons as to why there has been an increase in earmarking. In policy circles, earmarking in the United Nations Development System (UNDS) is seen as a response to unmet needs regarding the efficiency and results-orientation of UN organizations (Muttukumaru 2015; Reinsberg et al. 2015; Sridhar and Woods 2013; Tortora and

Steensen 2014). Donors earmark as they are unsatisfied with how organizations operate and feel they need to monitor closely the progress of projects and intervene occasionally (Weinlich et al. 2020). In 2018, a total of 79% of the UNDS revenue was earmarked, and with such high levels of earmarking, according to Otto Baumann (2020), this can create several issues such as high transaction costs, drive fragmentation, make the UN less strategic and impactful and overall, it can damage the multilateralism of the UN and its work. Fragmentation in the humanitarian sector is considered to be the norm as aid recipient countries have to deal with many humanitarian organizations that implement a wide variety of often uncoordinated projects (De Renzio). Earmarking has direct consequences on the effectiveness of aid by weakening the agency of developing countries and by encouraging short-term projects that tend to be focused on tangible, measurable results rather than tackling more complex long-term issues that hinder sustainable transformations (Otto Baumann 2020).

Private and Non-state Donors

When a crisis occurs, corporations and their foundations frequently provide cash donations, in-kind goods and access to critical infrastructure (White 2012). Although the engagement of the for-profit industry in humanitarian response and the international aid and development sector is not a new concept, it has been receiving increasing attention in recent times for a number of reasons (Huckel Schneider and Negin 2016).

First, there has been an increase in the number of for-profit organizations playing a role in development and humanitarianism and there have been increases in corporate social responsibility initiatives (Kent and Burke 2011; Kovács and Spens 2007; Weiss 2013; Zyck and Kent 2014). Second, there has been an increase in the scope of roles played by for-profit organizations in these areas (Huckel Schneider and Negin 2016), such as for-profit organizations developing specific emergency supply chain logistics (Cozzolino 2012). Third, there has been a change in the level of involvement of the private sector in the humanitarian field, with the emergence of for-profit organizations conducting, coordinating and responding to acute humanitarian emergencies as a core part of their business (Huckel Schneider and Negin 2016).

In terms of providing funding for humanitarian purposes, fostering economic development and shaping development policy, non-state donors are becoming increasingly important (Desai and Kharas 2008, 2018; Esser and Bench 2011; Metzger et al. 2010; Werker and Ahmed 2008). For example, from 2009 to 2017, the Bill and Melinda Gates Foundation contributed over US$26.1 billion—which is a tenth of the official aid budget of the United States, where the foundation is headquartered (OECD 2019).

With the growing frequency and severity of natural disasters and violent conflicts, the demand for additional financial resources in response to humanitarian crises is vastly increasing (Global Humanitarian Assistance 2018). Official donors are often unable to provide the funding needed to respond to humanitarian crises or are unwilling to (Fuchs and Öhler 2020). Becerra et al. (2014) and Becerra et al. (2015) show that in the aftermath of natural disasters, compared to the economic damages caused, official aid flows are low. For example, in the aftermath of the Indian Ocean earthquake and tsunami in 2004, the majority of humanitarian assistance that was provided originated from private sources (Kim et al. 2016), in fact, companies in the United States singlehandedly mobilized more than US$565 million (Thomas and Fritz 2006).

Private donors have often been perceived as being more needs-oriented than official donors (Fuchs and Öhler 2020). According Desai and Kharas (2008), "while official donor allocations are influenced by, among other things, political coalitions, policy concerns, and colonial ties, NGO allocations are assumed to be influenced by need." NGOs have often been criticized for being financially dependent on official donors which could undermine the autonomy of NGOs when it comes to allocating aid (Fuchs and Öhler 2020). According to Edwards and Hulme (1996), the relationship between NGOs and state agencies is "too close for comfort" with NGOs often becoming the "implementer of the policy agendas" on behalf of governments. Moreover, traditional donors may largely fund NGO projects that are located in countries they favor (Dreher et al. 2012).

Similarly, corporations may also provide humanitarian assistance to their government's preferred locations in order to obtain favors in return (Fuchs and Öhler 2020). For instance, Bertrand et al. (2020) describe corporate philanthropy as an alternative to political campaign contributions and lobbying activities for companies that are seeking to influence policies as corporate philanthropy can be used to gain favor with lawmakers. The size of the corporation and whether or not it has political ties matters as an empirical study conducted by Gao (2011) shows that "large firms and firms who have political ties donate a significant more to disaster relief than smaller firms and firms who do not have political ties." Corporate aid provided in response to humanitarian crises is on the rise (Fuchs and Öhler 2020). According to a study conducted by White (2012) investigating corporate responsiveness to natural disasters, it was found that during the last decade in terms of both scale and diversity, corporate engagement in natural disaster response has significantly increased and "today, it is a central component of the international response machinery and is becoming more and more important with each new disaster."

RISE IN INNOVATIVE HUMANITARIAN FINANCING

In order to address humanitarian issues, financing strategies need to look beyond traditional donors and consider alternative funding sources. According to the World Health Organization (2009), innovative financing mechanisms are defined as non-traditional uses of overseas development assistance, joint public–private mechanisms and flows that fundraise by tapping into new and diverse resources that deliver innovative or new financial solutions to humanitarian and/or development context.

The largest volume of international humanitarian aid tends to be directed to protracted crises (Spiegel et al. 2018); however, the majority of funding provided to countries that are in protracted crises still arrive on an annual basis rather than in multi-year grants (Development Initiatives 2017). Furthermore, nearly half of all international humanitarian aid, mainly by government donors from the Organization for Economic Cooperation and Development (OECD), was given to multilateral organizations, primarily organizations within the UN system (Spiegel et al. 2018). As funding from some government donors was reduced, the potential of funding from private sources continued to increase by 25% to USD$6.9 billion from sources such as individuals, trusts and foundations, and corporations (Development Initiatives 2017).

Beyond current reform efforts to try and make humanitarian funding faster, more consistent and more effective, there is also a sense that there should be a shift for donors to move from grant-making to using a wider range of financial instruments. According to (Tjønneland 2020), there is also an interest from traditional donors and foundations to "explore different uses of grant funding to attract greater capital input from investors. New partnerships and financial instruments from across the philanthropic–commercial spectrum could be used to address the challenges facing humanitarian financing." This has become to be known as innovative humanitarian financing. Not only does it provide opportunities but there are also significant challenges such as conflict of interests that could arise from the proliferation of private financing into the humanitarian sector.

REFERENCES

Abdi, M., Tani, S., Osman, N., Sadaat., S., and Jan, N. (2008). Local capacities for peace: Addressing land-based conflicts in Somaliland and Afghanistan. Academy for Peace and Development, Hargeisa, Somaliland. Retrieved from: https://media.africaportal.org/documents/local_capacities_for_peace_2008_EN.pdf.

Abel, G. J., Brottrager, M., Cuaresma, J. C., and Muttarak, R. (2019). Climate, conflict and forced migration. *Global Environmental Change*, 54, 239–249.

Aflaki, A., and Pedraza-Martinez, A. J. (2016). Humanitarian funding in a multi-donor market with donation uncertainty. *Production and Operations Management*, 25(7), 1274–1291.

Akresh, R., Bhalotra, S., Leone, M., and Osili, U. O. (2012). War and stature: Growing up during the Nigerian civil war. *American Economic Review*, 102(3), 273–277.

Al-Mahaidi, A. (2020). Financing Opportunities for Durable Solutions to Internal Displacement: Building on Current Thinking and Practice. *Refugee Survey Quarterly*, 39(4), 481–493.

Alderman, H., Hoddinott, J., and Kinsey, B. (2006). Long term consequences of early childhood malnutrition. *Oxford Economic Papers*, 58(3), 450–474.

Alesina, A., and Dollar, D. (2000). Who gives foreign aid to whom and why? *Journal of Economic Growth*, 5(1), 33–63.

Allansson, M., Melander, E., and Themnér, L. (2017). Organized violence, 1989–2016. *Journal of Peace Research*, 54(4), 574–587.

Annen, K., and Strickland, S. (2017). Global Samaritans? Donor election cycles and the allocation of humanitarian aid. *European Economic Review*, 96, 38–47.

Bakewell, O. (2000). Uncovering local perspectives on humanitarian assistance and its outcomes. *Disasters*, 24(2), 103–116.

Baumann, M. O. (2020). How earmarking has become self-perpetuating in United Nations development co-operation. *Development Policy Review*, 1–17.

Becerra, O., Cavallo, E., and Noy, I. (2014). Foreign aid in the aftermath of large natural disasters. *Review of Development Economics*, 18(3), 445–460.

Becerra, O., Cavallo, E., and Noy, I. (2015). Where is the money? Post-disaster foreign aid flows. *Environment and Development Economics*, 20(5), 561–586.

Bertrand, M., Bombardini, M., Fisman, R. J., and Trebbi, F. (2020). Tax-exempt lobbying: Corporate philanthropy as a tool for political influence. *American Economic Review*, 110(7), 2065–2102.

Bommer, C., Dreher, A., and Perez-Alvarez, M. (2018). Regional and ethnic favoritism in the allocation of humanitarian aid (CESifo Working Paper Series 7038). Munich, Germany: CESifo. Retrieved from: www.cesifo.org/DocDL/cesifo1_wp7038.pdf.

Brück, T., Dunker, K. M., Ferguson, N. T. N., Meysonnat A., and Nillesen, E. (2018). Determinants and dynamics of forced migration to Europe: Evidence from a 3-D model of flows and stocks. *IZA Discussion Papers*, 11834.

Brück, T., and d'Errico, M. (2019). Food security and violent conflict: Introduction to the special issue. *World Development*, 117, 167–171.

Brzoska, M. (2019). Understanding the disaster–migration–violent conflict nexus in a warming world: the importance of International Policy Interventions. *Social Sciences*, 8(6), 167.

Brzoska, M., and Fröhlich, C. (2016). Climate change, migration and violent conflict: vulnerabilities, pathways and adaptation strategies. *Migration and Development*, 5(2), 190–210.

Bulíř, A., and Hamann, A. J. (2003). Aid volatility: An empirical assessment. *IMF Staff Papers*, 50(1), 64–89.

Bundervoet, T., Verwimp, P., and Akresh, R. (2009). Health and civil war in rural Burundi. *Journal of Human Resources*, 44(2), 536–563.

Burde, D., Guven, O., Kelcey, J., Lahmann, H., and Al-Abbadi, K. (2015). What works to promote children's educational access, quality of learning, and wellbeing in crisis-affected contexts. Education Rigorous Literature Review, Department for International Development. London: Department for International Development.

Burnside, C., and Dollar, D. (2000). Aid, policies, and growth. *American Economic Review*, 90(4), 847–868.

Burrows, K., and Kinney, P. L. (2016). Exploring the climate change, migration and conflict nexus. *International Journal of Environmental Research and Public Health*, 13(4), 443: 1–17.

Cabot-Venton, C. (2013). *Value for money of multi-year approaches to humanitarian funding*. London: Department for International Development. Retrieved from: https://www.gov.uk/government/uploads/system/uploads/attachment_data/file/226161/VfM_of_Multi-year_Humanitarian_Funding_Report.pdf.

Cazabat, C., and Yasukawa, L. (2020). *Unveiling the Cost of Internal Displacement: 2020 Report*. Geneva: Internal Displacement Monitoring Centre.

Retrieved from: https://www.internal-displacement.org/sites/default/files/publications/documents/IDMC_CostEstimate_final.pdf.

Clemens, M., and Radelet, S. (2003). The millennium challenge account: How much is too much, how long is long enough? Working Paper No. 23, Centre for Global Development, Washington.

Collier, P., and Dollar, D. (2001). Can the world cut poverty in half? How policy reform and effective aid can meet the international development goals. *World Development*, 29(11), 1787–1802.

Collier, P., and Dollar, D. (2002). Aid allocation and poverty reduction. *European Economic Review*, 46(8), 1475–1500.

Contreras, S., Niles, S., Roudbari, S., Harrison, J., and Kaminsky, J. (2020). Bridging the praxis of hazards and development with resilience: A case study of an engineering education program. *International Journal of Disaster Risk Reduction*, 42, 101347.

Cozzolino, A. (2012). Humanitarian Logistics: Cross-sector Cooperation in Disaster Relief Management. Heiderlberg: Springer.

De Renzio, P. (2005). Scaling up versus absorptive capacity: Challenges and opportunities for reaching the MDGs in Africa. Overseas Development Institute.

Dehérez, D. (2009). The Scarcity of Land in Somalia. Natural Recourses and their Role in the Somali Conflict. Bonn International Center for Conversion, Occasional Paper III, Bonn.

Dellmuth, L. M., Bender, F. A. M., Jönsson, A. R., Rosvold, E. L., & von Uexkull, N. (2021). Humanitarian need drives multilateral disaster aid. *Proceedings of the National Academy of Sciences*, 118(4).

Desai, R. M., and Kharas, H. (2008). The California consensus: Can private aid end global poverty? *Survival*, 50(4), 155–168.

Desai, R. M., and Kharas, H. (2018). What motivates private foreign aid? Evidence from internet-based microlending. *International Studies Quarterly*, 62(3), 505–519.

Development Initiatives. (2017). Global Humanitarian Assistance Report 2017. Retrieved from: http://devinit.org/post/global-humanitarian-assistance-2017/.

Dreher, A., Nunnenkamp, P., Öhler, H., & Weisser, J. (2012). Financial dependence and aid allocation by Swiss NGOs: A panel tobit analysis. *Economic Development and Cultural Change*, 60(4), 829–867

Edwards, S., and van Wijnberg, S. (1989). Disequilibrium and structural adjustment. In: H. Chenery and T. N. Srinivasan (Eds.), *Handbook of development economics*. Amsterdam.

Edwards, M., & Hulme, D. (1996). Too close for comfort? The impact of official aid on nongovernmental organizations. *World Development*, 24(6), 961–973.

Esser, D. E., and Bench, K. K. (2011). Does global health funding respond to recipients' needs? Comparing public and private donors' allocations in 2005–2007. *World Development*, 39(8), 1271–1280.

Fabre. (2017). Multi-Year Humanitarian Funding. World Humanitarian Summit Putting Policy into Practice. The Commitments into Action Series. OECD Development Co-Operation Working Paper (OECD). Retrieved from: http://www.oecd.org/development/humanitarian-donors/docs/multiyearfunding.pdf.

FAO, NRC & OCHA. (2017). Living Up to the Promise of Multi-Year Humanitarian Financing. Retrieved from: https://reliefweb.int/report/world/living-promise-multi-year-humanitarianfinancing.

Faye, M., and Niehaus, P. (2012). Political aid cycles. *American Economic Review*, 102(7), 3516–3530.

Fink, G., and Redaelli, S. (2011). Determinants of international emergency aid, humanitarian need only? *World Development*, 39(5), 741–757.

Fuchs, A., and Öhler, H. (2020). Does private aid follow the flag? An empirical analysis of humanitarian assistance. The World Economy.

Gao, Y. (2011). Philanthropic disaster relief giving as a response to institutional pressure: Evidence from China. *Journal of Business Research*, 64(12), 1377–1382.

Gemmell, N., and McGillivray, M. (1998). Aid and tax instability and the government budget constraint in developing countries. Centre for Research in Economic Development and International Trade. Research Paper No. 98/1, University of Nottingham, Nottingham.

Global Humanitarian Assistance. (2018). *Global Humanitarian Assistance Report 2018*. Bristol, UK: Development Initiatives.

Hamro-Drotz, D. (2014). Livelihood Security: Climate Change, Migration and Conflict in the Sahel. Conflict-Sensitive Adaptation to Climate Change in Africa. United Nations Enviroment Programme.

Handmer, J., Honda, Y., Kundzewicz, Z., Arnell, N., Benito, G., Hatfield, J., ... Yamano, H. (2012). Changes in Impacts of Climate Extremes: Human Systems and Ecosystems. In: C. Field, V. Barros, T. Stocker, and Q. Dahe (Eds.), *Managing the risks of extreme events and disasters to advance climate change adaptation: Special report of the intergovernmental panel on climate change* (pp. 231–290). Cambridge: Cambridge University Press.

Hansen, H., and Tarp, F. (2000). Aid and Growth Regressions. Centre for Research in Economic Development and International Trade, University of Nottingham.

Harris, K., Keen, D., and Mitchell, T. (2013). *When disasters and conflicts collide. Improving links between disaster resilience and conflict prevention*. London, UK: Overseas Development Institute.

Heller, P. S., and Gupta, S. (2002). More Aid—Making It Work for the Poor. *World Economics*, 3(4), 131–146.

Hoeffler, A., and Outram, V. (2011). Need, merit, or self-interest—What determines the allocation of aid? *Review of Development Economics*, 15(2), 237–250.

Hoffmann, K., and Kirk, T. (2013). Public Authority and the Provision of Public Goods in Conflict-Affected and Transitioning Regions. Justice and Security Research Programme Paper 7, LSE. International Development Department, London.

Hoffman, P. J. (2016). Humanitarianism in treatment: Analyzing The World Humanitarian Summit. Future United Nations Development System (FUNDS) Briefing, 41. Retrieved from: https://futureun.org/media/arc hive1/briefings/FUNDS_Brief41_WorldHumanitarianSummit_August2016. pdf.

Huckel Schneider, C., and Negin, J. (2016). The for-profit sector in humanitarian response: Integrating ethical considerations in public policy decision making. *Medicine, Conflict and Survival*, 32(3), 184–202.

Ide, T., and Scheffran, J. (2014). On climate, conflict and cumulation: Suggestions for integrative cumulation of knowledge in the research on climate change and violent conflict. *Global Change, Peace and Security*, 26(3), 263–279.

Ide, T., Link, P. M., Scheffran, J., and Schilling, J. (2016). The climate-conflict Nexus: Pathways, regional links, and case studies. In: *Handbook on sustainability transition and sustainable peace* (pp. 285–304). Cham: Springer.

Intergovernmental Panel on Climate Change (IPCC). (2018). Summary for policymakers in Global Warming of 1.5°C. An IPCC Special Report on the Impacts of Global Warming of 1.5°C above Pre-Industrial Levels and Related Global Greenhouse Gas Emission Pathways, in the Context of Strengthening the Global Response to the Threat of Climate Change. Sustainable Development, and Efforts to Eradicate Poverty. Geneva, Switzerland: World Meteorological Organization.

Jahangiri, K., Izadkhah, Y. O., and Tabibi, S. J. (2011). A comparative study on community-based disaster management in selected countries and designing a model for Iran. *Disaster Prevention and Management: An International Journal*, 20(1), 82–94.

Jahre, M., and Heigh, I. (2008). Does the current constraints in funding promote failure in humanitarian supply chains? *Supply Chain Forum: An International Journal*, 9(2), 44–54.

Kent, R., and Burke, J. (2011). Commercial and humanitarian engagement in crisis contexts: Current trends, future drivers. Humanitarian Futures Programme, King's College London, UK.

Killick, T. (2004). The Case against Doubling Aid to Africa notes for House of Commons Presentation.

Kim, Y., Nunnenkamp, P., and Bagchi, C. (2016). The Indian Ocean tsunami and private donations to NGOs. *Disasters*, 40(4), 591–620.

Koubi, V., Böhmelt, T., Spilker, G., and Schaffer, L. (2018). The determinants of environmental migrants' conflict perception. *International Organization*, 72(4), 905–936.

Kovács, G., and Spens, K. M. (2007). Humanitarian logistics in disaster relief operations. *International Journal of Physical Distribution & Logistics Management*, 37(2), 99–114.

Lancaster, C. (2007). *Foreign Aid: Diplomacy, development, domestic politics*. Chicago: University of Chicago Press.

Lemaitre, J. (2018). Humanitarian aid and host state capacity: The challenges of the Norwegian Refugee Council in Colombia. *Third World Quarterly*, 39(3), 544–559.

Levin, V., and Dollar, D. (2005). The forgotten states: aid volumes and volatility in difficult partnership countries (1992–2002). Summary Paper for DAC Learning and Advisory Process on Difficult Partnerships, OECD-DAC, Geneva.

McGillivray, M. (2005). Aid Allocation and Fragile States. Background Paper for the Senior Level Forum on Development Effectiveness in Fragile States, 13–14 January, Lancaster House, London.

Metzger, L., Nunnenkamp, P., and Mahmoud, T. O. (2010). Is corporate aid targeted to poor and deserving countries? A case study of Nestlé's aid allocation. *World Development*, 38(3), 228–243.

Muttukumaru, R. (2015). The Funding and related practices of the UN Development system: ECOSOC Dialogue on the longer-term positioning of the UN Development system in the context of the post-2015 development agenda. https://www.un.org/en/ecosoc/qcpr/pdf/ie_muttukumaru_paper_funding.pdf.

Nunnenkamp, P., and Öhler, H. (2012). How to attract donations: The case of US NGOs in international development. *The Journal of Development Studies*, 48(10), 1522–1535.

OECD. (2019). Aid (ODA) by sector and donor [DAC5]. OECD.stat. Paris, France: Organisation for Economic Cooperation and Development. Retrieved from https://stats.oecd.org/.

Opdyke, A., Goldwyn, B., and Javernick-Will, A. (2021). Defining a humanitarian shelter and settlements research agenda. *International Journal of Disaster Risk Reduction*, 52, 101950.

Poole, L., and Culbert, V. (2019). Financing the Nexus: Gaps and Opportunities from a Field Perspective. FAO, NRC, UNDP. Retrieved from: https://www.nrc.no/resources/reports/financing-the-nexus-gaps-and-opportunities-from-a-field-perspective.

Raschky, P. A., and Schwindt, M. (2012). On the channel and type of aid: The case of international disaster assistance. *European Journal of Political Economy*, 28(1), 119–131.

Reinsberg, B., Eichenauer, V. Z., and Michaelowa, K. (2015). The rise of multi-bi aid and the proliferation of trust funds. In: M. B. Arvin and B. Lew (Eds.), *Handbook on the economics of foreign aid* (pp. 527–554). Edward Elgar Publishing Limited.

Rysaback-Smith, H. (2015). History and principles of humanitarian action. *Turkish Journal of Emergency Medicine*, 15, 5–7.

Safarpour, H. (2018). Donors management in disasters: Kermanshah earthquake experience. *Iranian Red Crescent Medical Journal*, 20(11), e84942.

Safarpour, H., Fooladlou, S., Safi-Keykaleh, M., Mousavipour, S., Pirani, D., Sahebi, A., Ghodsi, H., Farahi-Ashtiani, I., and Dehghani, A., (2020). Challenges and barriers of humanitarian aid management in 2017 Kermanshah earthquake: a qualitative study. *BMC Public Health*, 20, 1–10.

Scheffran, J., Brzoska, M., Kominek, J., Link, P., and Schilling, J. (2012). Climate Change and Violent Conflict. *Science(Washington)*, 336(6083), 869–871.

Schilling, J., Nash, S. L., Ide, T., Scheffran, J., Froese, R., and Von Prondzinski, P. (2017). Resilience and environmental security: towards joint application in peacebuilding. *Global Change, Peace & Security*, 29(2), 107–127.

Scott, R. (2014). Imagining more effective humanitarian aid: A donor perspective. Retrieved from: http://www.oecd.org/dac/Imagining%20More%20Effective%20Humanitarian%20Aid_October%202014.pdf.

Scott, R. (2015). Financing in Crisis? Making Humanitarian Finance Fit for the future. OECD Development Cooperation Working Paper 22. June 2015. Paris, France.

Siddiqi, A. (2018). Disasters in conflict areas: Finding the politics. *Disasters*, 42, S161–S172.

Singhal, S. (2019). Early life shocks and mental health: The long-term effect of war in Vietnam. *Journal of Development Economics*, 141, 102244.

Spiegel, P., Chanis, R., and Trujillo, A. (2018). Innovative health financing for refugees. *BMC Medicine*, 16(1), 1–10.

Sridhar, D., and Woods, N. (2013). Trojan multilateralism: Global cooperation in health. *Global Policy*, 4(4), 325–335.

Straus, S. (2012). Wars do end! Changing patterns of political violence in sub-Saharan Africa. *African Affairs*, 111(443), 179–201.

Stumpf, J., Guerrero-Garcia, S., Lamarche, J. B., Besiou, M., and Rafter, S. (2017). Supply chain expenditure & preparedness investment opportunities in the humanitarian context. Tech. rep., Action Contre la Faim—ACF France.

Retrieved from: https://www.actioncontrelafaim.org/wp-content/uploads/2018/05/ACF_Report_Supply-Chain-Exp-and-Inv.-Opportunities_2017 1124_Final.pdf.

The Statistical, Economic and Social Research and Training Centre for Islamic Countries (SESRIC). (2017). Humanitarian Crises in OIC Countries— Drivers, Impacts, Current Challenges, and Potential Remedies. Ankara, Turkey. Retrieved from: https://www.sesric.org/files/article/573.pdf.

Thomas, A., and Fritz, L. (2006). Disaster relief, Inc. *Harvard Business Review*, 84(11), 114–122.

Thulstrup, A. W., Habimana, D., Joshi, I., and Oduori, S. M. (2020). Uncovering the challenges of domestic energy access in the context of weather and climate extremes in Somalia. *Weather and Climate Extremes*, 27, 100185.

Tjønneland, E. (2020). Financing. In: *Humanitarianism: Keywords* (pp. 67–69). Leiden: Brill.

Tortora, P., and Steensen, S. (2014). Making earmarked funding more effective: Current practices and a way forward. Retrieved from: https://www.oecd.org/dac/aid-architecture/Multilateral%20Report%20N%201_2014.pdf.

Turrent, V., and Oketch, M. (2009). Financing Universal Primary Education: An Analysis of Official Development Assistance in Fragile States. *International Journal of Educational Development*, 29(4): 357–365.

Turrini, L., Besiou, M., Papies, D., and Meissner, J. (2020). The role of operational expenditures and misalignments in fundraising for international humanitarian aid. *Journal of Operations Management*, 66(4), 379–417.

UCDP (Uppsala Conflict Data Program). (2018). Retrieved from: http://ucdp.uu.se/.

UNESCO. (2011). *The Hidden Crisis: Armed Conflict and Education. Education for All (EFA) Global Monitoring Report*. Paris: UNESCO.

United Nations High Commissioner for Refugees (UNHCR). (2018). Global Trends. Forced Displacement in 2017.

United Nations High Commissioner for Refugees (UNHCR). (2019). Global Trends: Forced displacement in 2018.

United Nations High Commissioner for Refugees (UNHCR). (2020). Global Trends: Forced Displacement in 2019. Geneva: UNHCR. Retrieved from: https://www.unhcr.org/globaltrends2019/.

van Wassenhove, L. N. (2006). Humanitarian aid logistics: Supply chain management in high gear. *The Journal of the Operational Research Society*, 57(5), 475–489.

Vitoriano, B., Ortuño, M. T., Tirado, G., and Montero, J. (2011). A multi-criteria optimization model for humanitarian aid distribution. *Journal of Global Optimization*, 51(2), 189–208.

Wakolbinger, T., and Toyasaki, F. (2011). Impacts of funding systems on humanitarian operations. In: M. Christopher and P. Tatham (Eds.), *Humanitarian logistics: Meeting the challenge of preparing for and responding to disasters* (pp. 33–46). London, UK: Kogan Page.

Walt, G. (1994). *Health policy: An introduction to process and power*. London: Zed Books.

Wedge, J. (2008). Where peace begins: Education's role in conflict prevention and peace building. Retrieved from: http://reliefweb.int/sites/reliefweb.int/files/resources/A2A4C1378EF2807E4925757E001E42DE-200803_Where_Peace_Begins.pdf.

Weinlich, S., Baumann, M.-O., Lundsgaarde, E., and Wolff, P. (2020). Earmarking in the multilateral development system: Many shades of grey.

Weiss, T. G. (2013). *Humanitarian Business*. Wiley.

Werker, E. D., and Ahmed, F. Z. (2008). What do nongovernmental organizations do? *Journal of Economic Perspectives, 22*(2), 73–92.

White, S. (2012). Corporate engagement in natural disaster response: Piecing together the value chain. Center for Strategic and International Studies, Washington. Retrieved from: http://csis.org/files/publication/120117_White_CorporateEngagement_Web.pdf.

World Bank. (2020). Somalia. Retrieved from: https://data.worldbank.org/country/somalia?name_desc=false.

World Health Organization. (2009). Taskforce for Innovative International Financing for Health Systems. Working Group 2: Raising and Channelling Funds. Progress report to Taskforce. Retrieved from: http://www.who.int/pmnch/media/membernews/2009/20090319_tfwg2.pdf.

Zyck, S., & Kent, R. (2014). Humanitarian crises, emergency preparedness and response: The role of business and the private sector. In: *Humanitarian Policy Group*. London: Overseas Development Institute.

CHAPTER 3

Innovative Vaccine and Immunization Funding: Advance Market Commitments

INTRODUCTION

As defined by the World Health Organization (WHO), Universal Health Coverage (UHC) is "all individuals and communities having access to any health services they need, of sufficient quality to be effective, without suffering financial hardship" (Ifeagwu et al. 2021). Among other things, the United Nations (UN) Sustainable Development Goals commit to achieving UHC by 2030 whereby people all over the world should have access to quality health services they need—whether its promotive, preventive, curative, rehabilitative, or palliative healthcare—without enduring financial hardship (Kieny et al. 2018). Moreover, in 2019, UN Member States at the 74th United Nations General Assembly recommitted to achieving UHC for a healthier world by 2030 (Reddy et al. 2020). However, effective strategies to funding healthcare remain a challenge in low and middle-income countries (LMICs) and humanitarian contexts.

The global economic crisis adversely impacted development assistance for health (DAH) (Atun et al. 2017), going from a peak of US$38 billion in 2013, DAH fell to USD$36.3 billion in 2015 (Dieleman et al. 2016a). Moreover, funding channeled through more traditional sources of financing such as bilateral and multilateral agencies, followed a similar trajectory; UN agencies accounted for 27.6% of DAH in 1990 which

© The Author(s), under exclusive license to Springer Nature Switzerland AG 2021
M. Ahmed, *Innovative Humanitarian Financing*,
Palgrave Studies in Impact Finance,
https://doi.org/10.1007/978-3-030-83209-4_3

fell to 12.4% in 2015, and funding from development banks fell from 18.6% of DAH in 2000 to 8.6% in 2015 (Dieleman et al. 2016a). Furthermore, according to Dieleman et al. (2016b), the trend of declining DAH financing will likely continue in the future; this will therefore increase reliance on domestic and innovative financing sources in order to sustain and scale up health programs in LMICs (Atun et al. 2017).

Annually, over a hundred million people are becoming impoverished due to devastating health costs and in particular in LMICs (WHO 2017a); therefore, developing funding solutions to address these issues is key. Following the International Conference on Financing for Development that took place in Monterrey, Mexico, innovative financing gained prominence (Atun et al. 2012) as a mechanism to provide additional funding for global health (Atun et al. 2017). A High-Level Taskforce on Innovative Financing for Health Systems was launched to identify various innovative financing sources such as airline taxes, tobacco taxes, immunization bonds, advance market commitments and debt swaps (McCoy and Brikci 2010; Fryatt et al. 2010) as alternative funding sources to supplement DAH.

The aim of the chapter is to evaluate the effectiveness of innovative health financing sources such as advance market commitments, as viable funding mechanisms for global health initiatives such as vaccine and immunization programs. This chapter is organized as follows; the next section will critically review the literature on public private partnerships, philanthropy, global health partnerships and initiatives. Section three will present a case study of the Pneumococcal Vaccine Advance Market Commitment Program and in particular, the development of the 10-valent pneumococcal conjugate vaccine offering some lessons learnt from this case. Section four will conclude and provide recommendations for further study.

LITERATURE REVIEW
Public–Private Partnerships (PPPs)

There are extensive debates about the definition of public–private partnerships (PPPs). Whether PPPs need a definition and what constitutes as a PPP is debated with some scholars arguing that there needs to be a redefinition of PPPs (Khanom 2010). For example, Hodge and Greve (2007) state that since a large number of definitions of PPPs can be found, there needs to be a re-examination of the various meanings and definitions given to PPPs. However, according to William (1997), PPPs need no specific definition as the concept is assumed well-defined as there is a consensus about the general definition of PPPs being the cooperative activities between both the public and private sectors. Simply put, a PPP is considered as an institutionalized form of cooperation between public and private actors who work together toward a joint target (Nijkamp et al. 2002). Furthermore, the WHO define PPPs as a collaboration of both public and private sector actors within diverse arrangements that differ according to; participants, legal status, governance, management, policy setting, contributions and operational roles to achieve specific outcomes (Kraak et al. 2012).

According to Khanom (2010), PPPs were seen as an outcome of New Public Management (NPM) which shifted the focus of management from public service to service delivery, and since the 1980s, the reinvention of the role of government, market mechanisms, privatization, deregulation and the contestability in the provision of public good and services were "the keywords of NPM". Furthermore, the central focus of NPM was to reduce public sector expenditure and facilitate a voluntary engagement with the public sector with the aim to provide public goods and an allocation of responsibilities to the private sector (Mitchell-Weaver and Manning 1991). Then in the 1990s, PPPs were established as a key public policy tool across the world (Osborne 2000) and have since become widely popular in public sector management (Khanom 2010).

Although PPPs were originally treated as a subset of the privatization movement, there is a growing consensus that PPPs do not simply mean the privatization of public services or the introduction of market mechanisms (Jamali 2004). Instead, PPPs imply some sort of collaboration to pursue common goals while at the same time leveraging joint resources and capitalizing on the particular competencies and strengths of both public and private partners (Widdus 2001; Pongsiri 2002; Nijkamp et al. 2002). Furthermore, according to Austin (2000), partnership relationships can be defined by engagement level, resource investments, managerial complexity and the strategic value of the association to each partner's mission.

PPPs have been considered by some as representing a middle path between state capitalism and privatization (Leitch and Motion 2003); however, critics argue that privatization did not result in substantial reductions in national debts neither did the private sector demonstrate its superiority in running businesses that had provided the "philosophical underpinnings" of the privatization process (Broadbent and Laughlin 2003; Leitch and Motion 2003). Moreover, PPPs were often seen as a way of involving the private sector in projects that were of national interest while at the same time avoiding the problems associated with the extensive privatizations that occurred in the 1980s (Jamali 2004).

In developing a PPP, public sector agencies and private sector organizations can seek mutual advantages particularly when private organizations are characterized by mutual respect, fairness, trust and openness (Jamali 2004). As for public sector agencies, the main rewards for partnering with the private sector are to improve the performance of programs, improve efficiencies in cost, enhance provisions of services and appropriate allocation of risks and responsibilities (Pongsiri 2002). Therefore, the interests of private sector organizations are the expectation to have a better investment potential, to yield a reasonable profit and to have better opportunities to expand their business interests (Jamali 2004) also, a reasonable return on investment is an essential consideration (Scharle 2002). The respective roles of both the public and private sectors are not identical but more so complementary and while PPPs can provide a means to mutually capitalize the comparative advantages of public and private sectors, there are several issues that need to be taken into consideration when considering a PPP (Jamali 2004).

For instance, the government has to maintain its involvement and this is especially the case when accountability is critical, when cost-shifting presents problems, or the timeframe of the project is long, or the case when societal normative choices are more important than costs (Spackman 2002). PPPs should not imply less of a governmental role but rather a different role as the position and strength of the private sector requires more of a skilled government participation (Scharle 2002). According to Jamali (2004), although PPPs may offer several opportunities, they should not be treated as a panacea instead, they should be evaluated on their merits on a case-by-case basis and whether the elements of effective collaboration are found or can be nurtured.

In the context of developing countries, the proliferation of PPPs has been attributed to several reasons such as: the desire to improve public sector performance by utilizing innovative operation and maintenance methods; reducing and stabilizing the costs of providing services; ensuring compliance with environmental requirements to improve environmental protection; enhancing competition; and accessing private capital for infrastructure investments to reduce government budgetary constraints (Miller 2000; Savas 2000). Also, there have been several changes in overseas development assistance that have taken place at the intersection of public and private spheres (Moran and Stone 2016). A variety of factors such as the proliferation of PPPs, the increasing involvement of the private sector in development (Di Bella et al. 2013) and the increased involvement of non-state actors, have moved development assistance from its position as largely the preserve of bilateral and multilateral donors (Desai and Kharas 2014).

Moreover, the landscape of development has been altered due to changes in the political economy of development finance (Moran and Stone 2016) as there have been significant increases in private aid flows since 2000 (Desai and Kharas 2014). These changes, both structural and institutional, have provided a space for the return of the role of philanthropy and foundations as active participants in development aid (Moran and Stone 2016). In this context, the importance and influence of philanthropy has strengthened (Desai and Kharas 2014); this is due to the continued strains that have been placed on bilateral and multilateral aid systems which has led to the questioning of the efficacy of official approaches to aid financing (Easterly 2006).

Philanthropy

The concept of philanthropy has been the concern of philosophers since early human history (Machuca 2006), in its literal sense, the term philanthropy means friend-love of the human species ("philos of the anthropos") (Christou et al. 2019) and philanthropy can be described as the habit of doing good and can be synonymous with goodness (Sulek 2010). From a corporate perspective, increased interest in philanthropy could be a reflection of a wider cultural shift and realization of firms that pursuing profits should not be their only purpose (Collins 1994; Hu and Yoshikawa 2017).

In international development, philanthropy has long been a source of financing and private philanthropic organizations have played a significant role in key development sectors (Moran and Stone 2016). For example, philanthropic organizations such as the Ford Foundation and others such as the MacArthur Foundation financed non-governmental activity from the 1970s to support the construction of global civil society (Moran and Stone 2016). More recently, the notion of philanthropy has received a lot of interest, particularly in the humanitarian sector. However, there have been a few criticisms against this "new" form of philanthropy. Philanthropic organizations have been criticized for exercising an undue influence on setting social policy agendas and circumventing democratic forms of policy deliberation and accountability (Birn 2014; Garett 2012; Szlezak et al. 2010).

Overall, the role of philanthropy in the international development and humanitarian sectors is deeply rooted, and in many respects, its current influence in building transnational policy partnerships is not a novel concept (Moran and Stone 2016). New institutional mechanisms for delivering aid and product development in global health, in particular PPPs, can trace some of their origins to private foundations such as the Rockefeller Foundation and more recently, the Bill & Melinda Gates Foundation (BMGF) (Moran 2014). This is not surprising given that almost by definition, grant-making foundations must foster and catalyze partnerships such as PPPs (Moran and Stone 2016).

Other transnational policy partnerships have been established such as the Open Society Foundations network (funded by investor George Soros) that have focused on the dissemination of expertise to developing and transition countries (Stone 2013). According to Moran and Stone (2016), these various and distinct partnerships demonstrate that philanthropy, and in particular private foundations, can be seen as a re-emerging power in international development shaping policy. For instance, the Rockefeller Foundation was a significant player in the nascent field of international health, funding activities targeting communicable diseases such as hookworm, vaccine development and malaria (Moran and Stone 2016). The foundation was also instrumental in the expansion of the international health architecture which contributed to the creation of the League of Nations Health Organisation, and consequently its successor the WHO (Youde 2013). The foundations institutionalized approach, with its focus on market-based solutions and its cross-sectoral structure, introduced a market element to global health governance (Moran 2014).

Playing a similar role, the BMGF focused its resources on large-scale global funds and emerged as a major institutional player that both complemented and challenged the established global health architecture (Moran and Stone 2016). The early focus of the Gates family philanthropic interests was on education but it has since shifted its influence and interest to global health, and in particular vaccines, with the Gates' family making a US$750 million donation to kick-start the Global Alliance for

Vaccines and Immunisations Alliance, otherwise known as GAVI, shortly before the establishment of the foundation (Moran and Stone 2016). BMGF is widely accepted to be the singular most important actor in global health (McCoy et al. 2009a) and is one of the largest single sources of funding support to global health initiatives (Garett 2012; McGoey et al. 2011; Szlezak et al. 2010). The foundations activities are characterized by a focus on technical interventions (Moran 2011) and the activities of the BMGF demonstrate that grant-making foundations must partner with other agencies in order to achieve their goals as they are fundamentally constrained by their structure (Moran 2014).

According to the Foundation Centre (2012), in 2010, US foundations dispersed approximately US$4.3 billion to international grants which comprised 21% of all giving. Also, it has been found that the majority of US international grants pass through official intermediaries such as the World Bank, North American or European headquartered non-governmental organizations and multi-sectoral global funds such as GAVI (Moran and Stone 2016). GAVI has been considered as "one of the triumphs of global health efforts" (Buse and Tanaka 2011) and has come to symbolize global health partnerships.

Global Health Partnerships (GHPs)

In LMICs, the involvement of private actors in the provision of global health is not a recent phenomenon. However, what is considered relatively new in the rise of PPPs is the scale of the phenomenon and the significance that it has for the "on-going reconfiguration of the relation between the public and private sectors" (Languille 2017). Ruckert and Labonté (2014) argue that notwithstanding the proliferation of PPPs in global health, there is little clarity with regard to definition as there are a variety of terms used to describe different kinds of PPPs including Global Public Private Partnerships (GPPPs), Global Health Initiatives (GHIs), International Public Private Partnerships for Health (IPPPHs) and Global Health Partnerships (GHPs). According to the WHO, public–private alliances are considered as partnerships that "bring together a set of actors for the common goal of improving the health of a population through mutually agreed roles and principles" and are seen as important to achieving sustainable improvements in health on a global scale (Kickbusch and Quick 1998).

GHPs have become an important component of efforts to enhance global health results. This is partly due to the shift in global development and humanitarian assistance, with the creation of new aid mechanisms and global partnerships with mandates to achieve specific disease control targets (Brugha 2008). Also, the proliferation of new actors has also meant new challenges in effective global health coordination efforts (Brugha 2008) with some prominent GHPs operating in parallel to multilateral organizations directly competing for donor resources (Ruckert and Labonté 2014).

There are also other factors that explain the expansion of health partnerships. First, PPPs are perceived to be branches of neoliberal globalization and the ideological shift that it created (Buse and Walt 2000a, 2002). This is as in the first phase of the neoliberal era in the 1980s, there was a call for the retreat of the state and its downsizing through privatization, the emergence of PPPs coincided with the next stage of neoliberalism in the 1990s, when the role of the state was transformed to "correct market failures" and "enable the private sector to thrive" (Languille 2017). Second, the emergence of PPPs as a legitimate policy model for health provision is due to the power shift that occurred among multilateral organizations (Languille 2017). Since the 1980s, advocates for the virtues of the market such as the World Bank, the International Finance Corporation and the Organization for Economic Cooperation and Development, started to dominate UN agencies as the dominant players in global health policy arenas (Languille 2017; Zammit 2003). Third, health PPPs emerged as a response to the rise of new biotechnology; changes in intellectual property rights; the role of corporate social responsibility; epidemics such as AIDS and the market failures to develop medicines and vaccines for orphan diseases; and the competition between the WHO and UNICEF (Buse and Walt 2000a, 2002).

Numerous scholars argue the benefits of GHPs. For instance, Ruckert and Labonté (2014) argue that GHPs have helped highlight concerns of certain global health issues and "contributed to generating additional resources and facilitating access to medication for impoverished populations." Furthermore, GHPs not only raised awareness about pressing global health problems and put specific health issues at the forefront of national and international policy agendas, but they also helped to mobilize new funding commitments to tackle diseases and develop new health products (Ruckert and Labonté 2014). Caines and Lush (2004) suggest that GHPs have improved national health policymaking through institutional reforms and health systems strengthening; moreover, Druce and Harmer (2004) recognize the contributions of GHPs in establishing norms and standards in treatment protocols, technical management and financial strategies. Of the most important of GHPs contributions to global health is arguably research and development in neglected areas and the facilitation of access to vaccines (Kickbusch and Quick 1998).

However, there are many criticisms of GHPs. Some studies have found that many PPPs place great emphasis on private providers in the delivery of selective health interventions, where profit interests can reduce pro-poor targeting of such programs (Malmborg et al. 2006). Moreover, Ruckert and Labonté (2014) argue that GHPs have undermined efforts to better harmonize the provision of aid and align donor activities "skewing national priorities of recipient countries by imposing those of donor partners"; furthermore, this poor harmonization has resulted in duplication, waste and the emergence of parallel systems for healthcare service provision among GHPs (McKinsey 2005) and created little alignment between recipient countries' and GHP financial management systems (Casper 2004). In regard to the cost of essential medicines, requirements for specific bulk-buying through GHPs can reduce the public bulk-buying abilities of developing countries which can increase the cost of essential medicines (Carlson 2004), this therefore suggests that non-alignment between GHPs and aid recipient countries' existing mechanisms and policies could undermine the effectiveness of global health programs (Ruckert and Labonté 2014).

In regard to health partnerships targeting LMICs, according to Languille (2017), the literature can be divided into three main streams according to the type of PPP they study. The first stream of literature on health PPPs covers in-country agreements for the building and maintenance of health facilities and the provision of health services (Languille 2017). This stream corresponds directly more to the conventional approach of PPPs which has been widely studied in the context of developed countries and their healthcare systems (Acerete et al. 2011; Barlow et al. 2010) but there is a dearth of literature and research on LMICs (Languille 2017).

The second stream of literature on health PPPs focuses on "demand-side financing schemes" that subsidize the purchasing of healthcare services from accredited providers (Languille 2017). It includes "voucher schemes" where all or a portion of the health services costs are paid for certain groups such as those with lower income or women, and "cash transfers" that reimburse users for their consumption of healthcare services (Ensor 2004).

The third stream focuses on GHIs and in particular, the Global Fund and GAVI (Birn and Lexchin 2011; Grundy 2010; Muraskin 2002; Storeng 2014) where GHIs can be defined as "collaborative relationship[s] which transcend national boundaries and bring together at least three parties, among them a corporation (and/or industry association) and an intergovernmental organization, so as to achieve a shared health creating goal on the basis of a mutually agreed and explicitly defined division of labor" (Buse and Walt 2000a). Most of these GHIs have primarily focused on the development and supply of vaccines and drugs against infectious diseases (Languille 2017); however, there are a small number of GHIs that target non-communicable diseases and the social and economic factors that impact health (Buse and Walt 2000b).

Global Healthcare Initiatives (GHIs)

The global health landscape has been transformed, in the past 10 to 15 years, by the proliferation of coalitions (parallel and overlapping), alliances or partnerships that are working toward different goals (Rushton and Williams 2011). There has been a variety of GHIs focusing on specific diseases, selected interventions, commodities or services and work through joint decision-making from multiple stakeholders from the public and private sectors including: multilateral agencies; donor bodies; philanthropic foundations; and civil society (Buse and Harmer 2007; Reich 2002). These GHIs have challenged the authority of the WHO as the global health leader (Brown et al. 2006) and have been assuming major positions within global health policy networks (Lee and Goodman 2002). Moreover, these GHIs control financial resources for health (McCoy et al. 2009b; Ravishankar et al. 2009) and are supported by "new philanthropy" such as the BMGF (Storeng 2014).

There have been some positive arguments for GHIs such as their ability to get specific health issues onto national and international agendas, stimulate research and development and advance access to cost-effective healthcare interventions (Buse and Harmer 2007). According to a comprehensive review conducted by the WHO to assess the impact of GHIs on countries health systems, the WHO found that with some adjustments to the way in which they are operated, GHIs offer critical opportunities to improve the "efficiency, equity, value for money, and outcomes in global public health" (Samb et al. 2009). Critics of GHIs counter these arguments and state that while GHIs substantially increase the resources available to global health, they also reinforce a corporate approach to governance and restrict technical solutions to health (Birn 2005, 2009). Furthermore, high-volume global funds from such partnerships risk the disruption of the policy and planning processes of recipient countries by for example, diverting governments from joint efforts to strengthening health systems and by introducing the "re-verticalization" of planning, management, and monitoring and evaluation systems (Biesma et al. 2009; Oliveira-Cruz 2008). Also, there is a growing body of evidence that GHIs have a number of unintentional negative consequences on health systems (Biesma et al. 2009; McCoy 2009), in general, these arguments are part of a wider enduring tension on the history of the international health discourse between vertical programs, which are often focused on singular diseases, and horizontal approaches such as primary healthcare (Cueto 2004; Mills 1983, 2005). However, some scholars attribute the expansion of disease-specific initiatives to include support for health system strengthening a response to the growing pressure on prominent GHIs such as the Global Fund and GAVI (Hafner and Shiffman 2013; Marchal et al. 2009).

Case Study: The Pneumococcal Vaccine Advance Market Commitment Program

In 1974, the Expanded Programme for Immunization (EPI) was launched, and at that time, only 5% of children in low-income countries had access to immunization (Kim-Farley 1992); a few years later, the WHO established policies for immunization with the aim of immunizing more than 80% of the global birth cohort by 1990 (WHO 2011). Between the 1980s and 1990s, there were various innovations in the global vaccine markets and four trends in particular had a profound impact, these were (Gilchrist and Nanni 2013): (1) many high-income countries switched from the use of more traditional EPI vaccines to new or alternative vaccines; (2) high-income countries switched from the use of multi-dose vaccine presentations to single-dose vials and pre-filled syringes without preservatives (Milstien et al. 2005); (3) several high-income countries introduced vaccines against diseases and prepared for the introduction of more vaccines in late stage development; and (4) in high-income countries, the valence of pediatric vaccines increased from either monovalent or trivalent combinations to tetravalent, pentavalent and then hexavalent combinations.

Following the introduction of the EPI, global efforts to achieve universal immunization coverage have been increasing (Ikilezi et al. 2020). Moreover, the EPI initiative has resulted in immunization coverage increasing from 5 to 86%, and in some countries, it has reached up to 95% (WHO 2017b). A result of expanding immunization coverage for all children was "the globalization of the vaccine market" (Gilchrist and Nanni 2013). Vaccines are considered to be one of the most cost-effective and efficient health interventions; however, despite efforts to distribute vaccines to developing countries, throughout the 1990s, it became evident that the gap between high-income countries and LMICs was widening (Moran 2008). For LMICs, market failures for health products such as vaccines and other health products are created when there are mismatches between supply and demand, which leads to markets functioning sub-optimally (Barder et al. 2005; Batson 2005) and often constraining the implementation of public health programs (Malhame et al. 2019). To address these market shortcomings, there have been strategies that have been implemented through GAVI (Batson 2005; Gandhi 2015) that according to Malhame et al. (2019) have produced some tangible benefits.

In 2000, GAVI (formerly known as the Global Alliance for Vaccines and Immunization) was created as a PPP that aimed to improve access to new and underused vaccines for children living in developing countries. GAVI's principal aim is to increase the uptake of vaccines as well as reduce the historical delay of 15–20 years for new vaccines in order to reach developing countries (Gandhi 2015). Other initiatives followed such as the Global Immunization Vision and Strategy that was developed by the WHO and UNICEF in 2006 as well as Global Vaccine Action Plan (GVAP), a multi-sectoral initiative launched in 2012 (WHO 2013), among others.

The ability to secure continuous funding has long been a challenge for humanitarian organizations and development agencies. As such, GAVI helped pioneer two long-term innovative funding mechanisms: Advance Market Commitment (AMC) and the International Finance Facility for Immunization (IFFIm). The AMC and IFFIm programs enabled GAVI to plan its funding several years in advance with both being managed by the World Bank on behalf of GAVI, which depending on needs draws down on funding (Browne 2017). Between 1990 and 2016, DAH targeting immunization was USD$34.5 billion (Institute for Health Metrics and Evaluation 2018) and of this amount, USD$12.4 billion was channeled through GAVI from 2000 and USD$22.1 billion was disbursed through other DAH channels between 1990 and 2016 (Ikilezi et al. 2020). According to Ikilezi et al. (2020), in 2016 alone immunization DAH was estimated to be at USD$3.2 billion, of which USD$1.4 billion was from GAVI while USD$1.8 billion was from other channels.

GAVI

GAVI is a Geneva headquartered global health partnership which aims to increase access to immunization and strengthen health systems. Initial funding for GAVI was provided by the BMGF and it brings together, the WHO, UNICEF, the World Bank, donor governments, international development and finance organizations, the pharmaceutical industry and representatives from developing countries (Muraskin 2005). GAVI's mission is to save lives and protect people's health by increasing access to immunization in poorer countries (Kamya et al. 2017), and from inception, GAVI has engaged vaccine manufacturers as partners through representation on its governing body (Chee et al. 2008).

GAVI's mission statement is underpinned by four key objectives (Kassim and Abdullah 2017): (i) to accelerate equal access to and use of vaccines; (ii) to increase the effectiveness and efficiency of the delivery of immunization as an integrated part of strengthened healthcare systems; (iii) to enhance the sustainability of national immunization programs; and (iv) to shape the markets for vaccines and other immunization products. In terms of its organizational structure, GAVI is not an official organization as such, rather it is a PPP between relevant stakeholders in the vaccines and immunization field (Clemens et al. 2010).

GAVI was primarily created to increase access to new and underused vaccines in developing countries and to rapidly scale up those vaccines (Kallenberg 2016). It does that by inviting applications from governments of countries whose gross national income per capital is below GAVI's eligibility threshold, it then purchases vaccines through its procurement agencies such as UNICEF, and then provides vaccines to governments whose applications have been approved (Kallenberg 2016). During its first decade, GAVI received and dispersed an estimated US$4.5 billion on a "direct contribution basis" to procure and purchase vaccines for more than 70 developing countries in addition to extending grants from "cash windows" which aimed to support immunization services and strengthen health systems in developing countries (Clemens et al. 2010).

Between the years of 2000 to 2015, two-thirds of GAVI's funding came from donations by governments totaling an amount of US$11.6 billion, also, every five years, governments pledge to donate a certain amount and then commit to making regular payments to GAVI (Crocker-Buque and Mounier-Jack 2016). According to Zerhouni (2019), GAVI has taken a systematic approach to increasing access to vaccines and making them more affordable and also "defining an expanded scope of diseases for which vaccines are needed." GAVI has therefore lead global efforts to protect vulnerable groups such as children against diseases such as pneumococcal disease. Moreover, Gilchrist and Nanni (2013) argue that funding from GAVI and its partners created an entirely new market for vaccines that did not exist before.

Despite having a single-minded focus on the delivery of vaccines, from early 2005, the GAVI Board started to widen its support to health systems strengthening (Hill 2011; Naimoli 2009) and it is considered a key part of GAVI's mission. However, it has been argued that what is meant by health systems strengthening remains unclear as there are disputed understandings of the health systems agenda within the epistemic community around GAVI, which is a reflection of professional pressures and competing public health ideologies (Storeng 2014). Furthermore, unlike the WHO, GAVI does not engage in policy assistance or applied healthcare work (Stein and Sridhar 2018), rather it constitutes as a mechanism through which developing countries can aggregate their demand for vaccines in order to increase vaccine availability and lower prices (GAVI 2020a). However, although GHIs such as GAVI have developed a reputation for being more innovative and effective than the WHO, it is part of what has been called the "vilification of the public sector" by PPPs for health (Buse and Harmer 2007).

The Pneumococcal Vaccine Advance Market Commitment (AMC)

Vaccines are considered to be important for public health and an effective tool to reduce preventable deaths, and according to the WHO, vaccines save an estimated 2–3 million lives each year globally (WHO 2020a). However, the difficulty of meeting the health needs of developing countries is apparent in the market for vaccines as the delivery of vaccines in developing countries lags behind developed countries; moreover, the development of vaccines targeting diseases in developing countries has been disproportionately slow (Kremer et al. 2020b). Pharmaceutical companies have long been reluctant to invest in manufacturing new vaccines for the developing world due to a perceived lack of earning attractive returns (Snyder et al. 2011). To overcome this issue and to stimulate more investments, the AMC was created to guarantee vaccine manufacturers a long-term market (Snyder et al. 2011). Leveraging the AMC mechanism enabled GAVI to encourage new investments from manufacturers in addition to the introduction of additional manufacturers, while achieving and maintaining optimal pricing, quality and reliable supply (Zerhouni 2019).

One of the top infectious disease killer of children under the age of five globally is pneumonia, with most mortality in LMICs (UNICEF 2019). The WHO estimates that 5.8 million children (under the age of five) die each year, with an estimated 300,000 deaths being caused by pneumococcal infections (WHO 2019). The incidence of invasive pneumococcal disease (IPD) varies by country, and however, it has been reported to be 5-416/100,000 in developing countries; furthermore, an estimated 75% of IPD cases occur in children under the age of two (Maimaiti et al. 2013). The vaccination of infants is considered to be the most effective way to prevent infections and "reduce the burden, mortality and sequelae in children (direct effect) and adult populations (indirect effect)" (Suwantika et al. 2020).

In 2009, the Governments of Canada, Italy, Norway, Russia and the United Kingdom in addition to the BMGF funded and launched the AMC to reduce the impact of pneumococcal disease on children in LMICs (GAVI 2017) under the supervision of GAVI. The AMC was designed to fund the research and development of new vaccines by pharmaceutical companies where the funds were guaranteed against market failure (Browne 2017). Furthermore, the purpose of the AMC was to incentivize the manufacturers of vaccines to develop vaccines to the needs of those in LMICs as well as scale manufacturing of the new vaccines "by guaranteeing an initial purchase price and quantity of vaccines for purchase" (Reddy et al. 2020). With the pre-established commitment for volumes and predictable pricing, the pneumococcal AMC was able to draw in manufacturers in order to develop vaccines that might not be available otherwise (Reddy et al. 2020).

The initial idea to use an AMC to fund vaccine purchases was originally proposed by Kremer and Glennerster (2004) building upon ideas previously conceptualized by Kremer (2000a, 2000b). As pharmaceutical firms invest in research and development, donors pledge a fund to subsidize the initial purchases of newly developed vaccines above and beyond unit production costs (Kremer et al. 2020b). This assures firms that they will be able to recoup the large investments required to develop a new vaccine in developing countries, as well as overcome concerns regarding supplying vaccines to consumers in developing countries that may lack sufficient income to demand a high quantity of vaccines on their own, and concerns of humanitarian organizations using their bargaining power to "hold up" the pharmaceutical firms' investments, pushing the price toward unit production cost in ex post negotiations (Kremer et al. 2020b).

Pneumococcal conjugate vaccines (PCVs) have helped reduce severe childhood pneumonia and other pneumococcal diseases such as meningitis and sepsis since first licensure in 2000 (Roth et al. 2018; Wahl et al. 2018). The Gates Foundation and five countries pledged USD$1.5 billion toward a pilot AMC that targeted a PCV in 2007 as during that time, in developing countries, there was an existing PCV covering disease strain; however, in developing countries, PCVs covering the strains were in late-stage clinic trials (Kremer et al. 2020b). Thus far, three PCVs have been licensed to immunize children against pneumococcal disease (O'Brien 2017): (i) PCV7: a 7-valent PCV which is no longer available; (ii) PCV10: a 10-valent PCV; and (iii) PCV13: a 13-valent PCV.

The AMC called for firms to compete for ten-year supply contracts at a capped price of USD$3.50 per dose; therefore, a firm competing to supply XX million annual doses ($XX/200$ of the projected 200 million annual need) would need to secure an $XX/200$ share of the USD$1.5 billion AMC fund, which is paid out as a per dose subsidy for initial purchases (Kremer et al. 2020b). The AMC covered the 74 countries that were below the income threshold for GAVI eligibility and the country co-payments were set according to GAVI rules (Kremer et al. 2020b).

In October 2009, pharmaceutical companies GlaxoSmithKline, Pfizer (which by then had acquired Wyeth), the Serum Institute of India and Panacea Biotech filed for expressions of interest for the AMC—they expressed an interest to participate in such an agreement on a nonbinding basis (Snyder et al. 2011), GAVI then used its forecasts of demand for pneumococcal vaccines to set a goal of providing 200 million annual doses of the vaccines by 2015 (GAVI 2011). In 2010 GAVI set the first tender for 60 million doses (Snyder et al. 2011), and in March 2010, Glaxo-SmithKline and Pfizer both agreed to supply thirty million doses of their second-generation pneumococcal vaccines for ten years on an annual basis through the AMC from 2013 onwards (GAVI 2011). Overtime, GAVI issued more tenders sometimes outpacing country demand and in each case the pharmaceutical firms expanded their supply commitments in line with the tenders (Kremer et al. 2020b).

By 2016, PCV was distributed in 60 of the 73 GAVI eligible countries and annual distribution exceeded 160 million doses, which was enough to immunize over 50 million children per year and by 2018, nearly half of the target child population in GAVI countries were covered (Kremer et al. 2020b). For AMC, disbursements rapidly grew from USD$42.9 million in 2010 to USD$215 million in 2012 and dropped to USD$203 million by 2013 (Atun et al. 2017). Revenues also grew rapidly from USD$51.2 million in 2008 to USD$173.1 million in 2009, and fluctuated between USD$173.1 million and USD$114.1 million in 2010 through 2013, and a total of USD$804.8 million was generated between 2011 and 2015 (Atun et al. 2017).

In regard to cost-effectiveness, according to Tasslimi et al. (2011), the initial PCV rollout was highly cost effective. Moreover, estimates suggest that the introduction of PCVs saved 700,000 lives at a highly favorable cost (Kremer et al. 2020b). However, Arie (2011) argues that although GAVI has been successful in accelerating the introduction of new life-saving vaccines in LMICs, the alliance has been criticized for not reducing the price of vaccines to affordable levels.

10-Valent Pneumococcal Conjugate Vaccine (PNEUMOSIL®)

As each pharmaceutical companies' commitment was 15% of GAVI's 200 million dose target, GlaxoSmithKline and Pfizer were each allocated 15% of the USD$1.5 billion fund, equivalent to USD$225 million, in addition, the companies agreed to provide seven million doses in 2010, 24 million in 2011 and 20 million in 2012 to countries that were eligible for GAVI assistance (Snyder et al. 2011). Given only 30% of the USD$1.5 billion fund was allocated to GlaxoSmithKline and Pfizer, 70% remained in reserve. The reserve funds could therefore be used to secure commitments from the participating vaccine manufacturers or new suppliers, in order to meet more of GAVI's 200 million doses target in the future (Snyder et al. 2011).

The agreements with the AMC obliged vaccine manufacturers to supply whatever the amount was demanded up to the supply commitment; however, GAVI has the option (not the obligation) to purchase any amount up to the supply commitment and therefore manufacturers bear risk if the forecasted demand does not materialize (Snyder et al. 2011). To offset the risk, GAVI's procurement agent UNICEF, regardless of whether demand materialized, agreed to purchase a minimum of 20% of GlaxoSmithKline's committed supply in the first year, 15% in the second year and 10% in the third year (GAVI 2011).

With funding from BMGF, the Serum Institute of India, Pvt. Ltd. (SIIPL) and PATH initiated the development of a PCV (Alderson et al. 2021). Over a decade later, the 10-valent PCV—PNEUMOSIL was produced which is WHO prequalified, AMC approved and licensed in India (WHO 2020b). Furthermore, SIPL became the first member of the Developing Countries Vaccine Manufacturers Network to manufacture a PCV prequalified by WHO (2020c) in 2019 and the third vaccine to be qualified for the AMC program.

Using novel conjugation technology, the vaccine protects against ten serotypes that are most likely to cause disease in Asia and Africa (where the disease is of greater burden), which makes it more advantageous than previous PCV vaccines (Hayman and Pagliusi 2020). PNEUMOSIL's LMIC price is US$2 per dose which is a reduction of over 30% from the GAVI supported price for other PCVs and significantly less than non-GAVI prices (GAVI 2020b). Furthermore, it was the first PCV designed to target serotypes most likely to cause IPD in the highest burden regions of Africa and Asia as well as in Latin America and the Caribbean (Johnson et al. 2010).

In regard to clinical development, PNEUMOSIL's strategy evolved overtime. Initially, SIPL and PATH planned to undertake clinical trials in India but they jointly determined that the easiest way to receive WHO prequalification, and market authorization in India, was by conducting clinical trials and development in both India and Africa (Alderson et al. 2021). In addition, it would also enable them to generate data and findings from a more diverse sample set. As such, SIPL led and funded the Indian studies for market authorization while PATH sponsored the African studies component with a grant from BMGF, which was required for WHO prequalification (Alderson et al. 2021).

PNEUMOSIL is now an option for LMICs to consider for introducing PCVs or switching programs. Its market entry marks an important milestone as it not only provides an affordable option but it is also adding a new PCV supplier (Alderson et al. 2021) ending the duopoly given that in 2018, there were only two manufacturers who had WHO prequalified PCVs, which provided coverage for 70–82% of childhood pneumococcal disease cases in Africa, Asia, Latin America, the Caribbean and the Pacific (UNICEF 2018). Furthermore, for GAVI and the countries it supports, PNEUMOSIL's 30% savings over other PCVs could enable public funds to go further to support other health priorities and tackle pneumococcal disease (Alderson et al. 2021).

Lessons Learned

There are a few lessons that can be learnt from PNEUMOSIL's development among them being factors related to: serotype selection and cost-effectiveness, product diversity and industry competition, market development and the financialization of healthcare.

(i) Serotype selection and cost-effectiveness

Relatively high PCV prices have made access for many countries difficult and only possible for LMICs eligible for GAVI financial support (GAVI 2020b) as PCVs are extreme complex to develop and manufacture. Reducing prices of new vaccines by manufacturing more cost-effective vaccines could therefore help GAVI in its immunization efforts. The introduction of PNEUMOSIL is an important step toward enhancing the cost-effectiveness of PCVs as well as boosting their supply in LMICs.

In order to reduce PCV prices and reduce cost of goods, optimizing development and manufacturing of the vaccine through serotype selection is key (Alderson et al. 2021). As such, in developing the PCV, the Pneumococcal Global Serotype Project and other disease burden data were used to help prioritize serotypes causing the highest pediatric IPD incidences in Africa, Asia, Latin America and the Caribbean (Alderson et al. 2021) and eventually ten serotypes were chosen for optimal balance of cost, coverage and competitiveness with existing PCVs (Johnson et al. 2010). Serotypes 6A and 19A emerged as significant causes of IPD in target regions; therefore, they were critical inclusions to ensure that for serotype coverage, PNEUMOSIL matched or exceeded other PCVs (Alderson et al. 2021).

Therefore, PNEUMOSIL was significantly cheaper than previous PCVs and managed to achieve a 30% saving compared to other PCVs. PNEUMOSIL was able to achieve a lower price point because as serotypes vary by region (which adds expense, risk and time to the development and manufacturing process), only serotypes most relevant for LMICs were included while those less applicable were omitted, which proved valuable to the overall coverage, performance and cost of PNEUMOSIL (Alderson et al. 2021).

(ii) Product diversity and industry competition

Another lesson that can be learnt from the development of PNEUMOSIL is the importance of product diversity and industry competition in order to increase vaccine supply and drive down prices.

In 2017, UNICEF procured 157 million doses of PCV with estimates that demand could further increase to 258 million doses by 2027, with one billion doses needed to immunize 361 million children between 2021 and 2027 (Hayman and Pagliusi 2020). The significant increase in demand prompted a call for a greater number of manufactures to supply more vaccines. This provided a powerful incentive to reduce the global burden of pneumococcal infections by enhancing competitiveness through product innovation and more manufacturers (Hayman and Pagliusi 2020).

The introduction of PNEUMOSIL therefore added more competition to the PCV suppliers market. It ended the "decade-long drought in new PCV suppliers" (Alderson et al. 2021) and could help reshape the market in the long-term by providing a more cost-effective vaccine.

(iii) Market development

The AMC was created to speed up the manufacturing and access of PCVs for LMICs. However, in terms of the market for PCVs, it still remains somewhat of a monopoly as after the creation of PNEUMOSIL, there are only three vaccines that have been qualified for the AMC program so far. As the new market entrant, SIPL is expected to participate in tenders for the remaining uncommitted AMC funds (Kremer et al. 2020a).

Going forward, a key issue for AMC will be whether to divide the AMC among multiple suppliers and reserve tender for entrants in the future (Kremer et al. 2020a) or concentrate invectives on a single supplier as initially proposed in Kremer and Glennerster (2004). Although this is unclear at the moment, there are clinical trials already underway in China and Russia by Walvax and Beijing Minhai, and Nanolek, respectively, who reportedly have undergone PCV Phase III clinical trials (Hayman and Pagliusi 2020).

(iv) Financialization of Healthcare

The rise of financial mechanisms such as GAVI can be seen as part of a wider trend of the financialization of global healthcare which has raised several concerns. Scholars from public health, international relations and international development theorists are primarily concerned with the ethical and political implications of having private sector involvement (Bull et al. 2004). For example, Richter (2004) highlights the depoliticizing nature of the discussion on partnerships whereby these arrangements are represented as a kind of "win-win-win" situation in which all parties mutually beneficially gain.

While disease specific and technical interventions may be beneficial in responding to humanitarian crises, to what extent they tackle the challenges facing health systems in developing countries is highly debated. One of the main criticisms of GAVI is it distorts the health sector priorities of developing countries as critics allege the vertical targeting of diseases and immunization cause distortions in weak and underfunded health systems (Garrett 2007). Furthermore, GHIs have been blamed for "cherry picking" certain groups that are easier to reach and therefore contributing to the growing inequality, also, a vast amount of the literature shows that GHIs are not entirely aligned with the national strategic plans of the countries' they are operating in and often inflict preconceived ideas (Hanefeld 2008; Cruz and McPake 2011).

According to Storeng (2014), there is an impetus for quick gains using mechanisms that bypass systemic issues such as under resourced public healthcare systems. Moreover, the shift to deliver development through multilateralism has been criticized for a number of reasons such as the undermining of country ownership by amplifying voices of donors, including non-traditional donors such as philanthropic foundations and the private sector (Sridhar and Woods 2013; Ooms 2007). Hunter and Murray (2019) argue that private investment in healthcare delivery and financing is now being presented as "the only solution for addressing geographic gaps in healthcare provision, high mortality and catastrophic out-of-pocket expenditures." Ryman et al. (2008) state that although vaccine coverage can reduce the occurrence of and mortality from various diseases, immunization service delivery needs to be strengthened and better integrated into general health services. While there may be short-term advances in certain diseases or vaccination coverage, it runs the risk of fragmenting primary health services (Sridhar and Tamashiro 2009).

Mechanisms such as AMCs legally bind governments to purchase vaccines for developing countries while vaccine manufacturers are subsidized by donors (Graham 2019); however, they have a track record of unevenly financializing private industry while health systems and services targeting the poor in general benefit little, despite the humanitarian value of initial public investments (Light 2005, 2009; Miraftab 2004; Schwalbe and El-Ziq 2010). For instance, through funding commitments, AMCs can provide predictable funding; however, AMC does not provide resources to LMICs to setup programs to help sell and distribute new products (Ketkar 2014) therefore in this respect, its role is fairly limited.

Conclusion

While there has been a proliferation of PPPs operating in the health sector throughout the first two decades of the twenty-first century (Widdus 2001), very few have focused on or worked in, vaccine development or distribution (Santos Rutschman 2020) with the exception of GAVI. Funding from GAVI and its partners created an entirely new market for vaccines that did not exist before (Gilchrist and Nanni 2013) as GAVI deals with the procurement and large-scale delivery of vaccines in mostly developing countries. For the past two decades, much of the vaccine procurement for children has been through GAVI, which has introduced 496 vaccines and contributed to the vaccination of 760 million children globally (GAVI 2020c). The creation of the AMC was part of an initiative to stimulate research, development and investment in vaccines for diseases such as pneumonia.

This chapter analyzes of the effectiveness of AMCs in incentivizing investment in vaccines for LMICs through a case study of the pneumococcal vaccine (PCV) and in particular, the development of the 10-valent PCV—PNEUMOSIL. A few lessons can be learnt from the case such as the importance of manufacturing strategy and optimization through serotype selection, the importance of product diversity and industry competition in order to increase vaccine supply and drive down prices and the growing concerns regarding the financialization of global healthcare.

In recent times, the role of AMCs has grown in importance and relevance. In the midst of the COVID-19 pandemic, there were calls to launch an AMC to accelerate the development and manufacturing of vaccines (Athey et al. 2020). As such GAVI announced a USD$2 billion international funding mechanism, COVAX, that contained a component labeled an AMC that offered vaccine manufacturers volume guaranteed in advance of licensure to increase supply to developing countries (GAVI 2020d). For instance, under Operation Warp Speed, the US government offered a variety of contracts to manufacturers and a deal was created with Pfizer committing to a USD$2 billion payment contingent on licensure or emergency use authorization of 100 million doses (U.S. Health and Human Services 2020).

- Further Research Recommendations

In regard to global public health interventions, there is an increased interest in the prevention of diseases through vaccinations in LMICs. Therefore, going forward, further research and evaluations of AMCs and AMC pilots such as PCVs, and their effectiveness and applicability to other interventions in LMICs, are much-needed. In particular, further studies should be conducted assessing and evaluating the cost-effectiveness, design, processes and impact of the various PCVs and how the pilots compare to each other in those aspects.

Notwithstanding the AMC, GAVI has multiple sources of innovative financing such as the GAVI matching fund and the IFFIm, and as a PPP, GAVI is largely funded through the IFFIm which is considered to be an innovative way of using overseas development assistance (ODA). Through the IFFIm, "vaccine bonds" are issued on the international capital markets to convert long-term donor pledges to immediately available cash resources (Atun et al. 2012). Going forward, further research should be conducted on the role capital markets play in raising and mobilizing funds to address humanitarian needs and in particular, how global capital markets are used to convert long-term donor contributions into current or "frontloaded" cash through the issuance of "vaccine bonds."

Moreover, further research should also be conducted on the use of Islamic capital markets by the IFFIm. More specifically, a comparative study including an analysis of conventional vaccine bonds and sukuk (Islamic bonds) issued by IFFIm could help develop a better understanding of the role Islamic capital markets play in raising and channeling funds for humanitarian purposes in comparison with conventional capital markets. The evaluation of these financial instruments in regard to their involvement to GAVI is important to understand how they impact the availability and predictability of donor funding for immunization.

References

Acerete, B., Stafford, A., and Stapleton, P. (2011). Spanish healthcare public private partnerships: The 'Alzira model'. *Critical Perspectives on Accounting*, 22(6), 533–549.

Alderson, M. R., Sethna, V., Newhouse, L. C., Lamola, S., and Dhere, R. (2021). Development strategy and lessons learned for a 10-valent pneumococcal conjugate vaccine (PNEUMOSIL®). Human Vaccines & Immunotherapeutics, 1–8.

Arie, S. (2011). How should GAVI build on its success? *BMJ*, 343.

Athey, S., Kremer, M., Snyder, C., and Tabarrok, A. (2020). New York Times. In the race for a coronavirus vaccine, we must go big. really, really big. Retrieved from: https://www.nytimes.com/2020/05/04/opinion/cor onavirus-vaccine.html.

Atun, R., Knaul, F. M., Akachi, Y., and Frenk, J. (2012). Innovative financing for health: what is truly innovative? *The Lancet*, 380(9858), 2044–2049.

Atun, R., Silva, S., and Knaul, F. M. (2017). Innovative financing instruments for global health 2002–15: a systematic analysis. *The Lancet Global Health*, 5(7), e720–e726.

Austin, J. E. (2000). Strategic collaboration between nonprofits and businesses. *Nonprofit Voluntary Sector, Quarterly*, 29, 69–97.

Barder, O., Kremer, M., and Levine, R. (2005). Answering concerns about Making Markets for Vaccines. Geneva: Commission on Intellectual Property Rights, Innovation and Public Health (CIPIH). Retrieved from: www.who.int/intellectualproperty/submissions/BarderSubmission.pdf.

Barlow, J., Roehrich, J. K., and Wright, S. (2010). De facto privatization or a renewed role for the EU? Paying for Europe's healthcare infrastructure in a recession. *Journal of the Royal Society of Medicine*, 103(2), 51–55.

Batson, A. (2005). The problems and promise of vaccine markets in developing countries. *Health Affairs*, 24(3), 690–693.

Biesma, R. G., Brugha, R., Harmer, A., Walsh, A., Spicer, N., and Walt, G. (2009). The effects of global health initiatives on country health systems: A review of the evidence from HIV/AIDS control. *Health Policy and Planning*, 24, 239–252.

Birn, A. E. (2005). Gates's grandest challenge: Transcending technology as public health ideology. *The Lancet*, 366, 514–519.

Birn, A. E. (2009). The stages of international (global) health: Histories of success or successes of history? *Global Public Health*, 4(1), 50–68.

Birn, A. E. (2014). Philanthrocapitalism, past and present: The Rockefeller Foundation, the Gates Foundation, and the setting(s) of the international/global health agenda. *Hypothesis*, 12(1), 1–27.

Birn, A. E., and Lexchin, J. (2011). Beyond patents: The GAVI Alliance, AMCs and improving immunization coverage through public sector vaccine production in the global south. *Human Vaccines*, 7(3), 291–292.

Broadbent, J., and Laughlin, R. (2003). Public private partnerships: an introduction. *Accounting, Auditing & Accountability Journal*, 16(3), 332–341.

Brown, T. M., Cueto, M., and Fee, E. (2006). The World Health Organization and the transition from "international" to "global" public health. *American Journal of Public Health*, 96(1), 62–72.

Browne, S. (2017). Vertical funds: New forms of multilateralism. *Global Policy*, 8, 36–45.

Brugha, R. (2008). Global Health Initiatives and Public Health Policy. In: *International Encyclopedia of Public Health*. Oxford: Academic Press.

Bull, B., Bøås, M., and McNeill, D. (2004). Private sector influence in the multilateral system: A changing structure of world governance? *Global Governance: A Review of Multilateralism and International Organizations*, 10(4), 481–498.

Buse, K., and Harmer, A. M. (2007). Seven habits of highly effective global public-private health partnerships: Practice and potential. *Social Science & Medicine*, 64, 259–271.

Buse, K., and Tanaka, S. (2011). Global Public-Private Health Partnerships: Lessons learned from ten years of experience and evaluation. *International Dental Journal*, 61, 2–10.

Buse, K., and Walt, G. (2000a). Global public–private partnerships: Part I A new development in health? *Bulletin of the World Health Organization*, 78(4), 549–561.

Buse, K., and Walt, G. (2000b). Global public–private partnerships: Part II—What are the health issues for global governance? *Bulletin of the World Health Organization*, 78(5), 699–709.

Buse, K., and Walt, G. (2002). The world health organization and global public-private health partnerships: In search of 'good' global health governance. In: R. Reich (Eds.), *Public-private partnerships for public health* (pp. 169–195). Cambridge, MA: Harvard University Press.

Caines, K., and L. Lush. (2004). *Impact of public-private partnerships addressing access to pharmaceuticals in selected low and middle-income countries: A Synthesis Report from Studies in Botswana, Sri Lanka, Uganda, and Zambia.* Geneva: Initiative on Public-Private Partnerships for Health.

Carlson, C. (2004). Assessing the Impact of Global Health Partnerships: Country Case Study Report (India, Sierra Leone, Uganda) (GHP Study Paper 7). London: Department for International Development Health Systems Resource Centre.

Casper, T. (2004). Updated discussion paper on the core business model of a mature global fund. Paper presented at the ninth board meeting of the Global Fund, Arusha.

Chee, G., Molldrem, V., Hsi, N., and Chankova, S. (2008). *Evaluation of the GAVI Phase 1 Performance (2000–2005).* Bethesda, MD: Abt Associates Inc.

Christou, P., Hadjielias, E., and Farmaki, A. (2019). Reconnaissance of philanthropy. *Annals of Tourism Research*, 78, 102749.

Clemens, J., Holmgren, J., Kaufmann, S. H., and Mantovani, A. (2010). Ten years of the Global Alliance for Vaccines and Immunization: Challenges and progress. *Nature Immunology*, 11(12), 1069–1072.

Collins, M. (1994). Global corporate philanthropy and relationship marketing. *European Management Journal*, 12(2), 226–233.

Crocker-Buque, T., and Mounier-Jack, S. (2016). The International Finance Facility for Immunisation: Stakeholders' perspectives. *Bulletin of the World Health Organization*, 94(9), 687.

Cruz, V. O., and McPake, B. (2011). Global Health Initiatives and aid effectiveness: Insights from a Ugandan case study. *Globalization and Health*, 7, 20.

Cueto, M. (2004). The origins of primary health care and selective primary health care. *American Journal of Public Health*, 94, 1864–1874.

Desai, R., and Kharas, H. (2014). The new global landscape of poverty alleviation and development: Foundations, NGOs, social media, and other private sector institutions. In: *Human dignity and the future of global institutions.* Washington, DC: Georgetown University Press.

Di Bella, J., Grant, A., Kindornay, S., and Tissot, S. (2013). *Mapping private sector engagements in development cooperation.* Ottawa: North-South Institute.

Dieleman, J.L., Schneider, M.T., Haakenstad, A., Singh, L., Sadat, N., Birger, M., Reynolds, A., Templin, T., Hamavid, H., Chapin, A., and Murray, C.J. (2016a). Development assistance for health: past trends, associations, and the future of international financial flows for health. *The Lancet*, 387(10037), 2536–2544.

Dieleman, J., Murray, C. L. J., and Haakenstad, A. (2016b). Financing Global Health 2015: Development assistance steady on the path to new Global Goals. Institute for Health Metrics and Evaluation. Retrieved from: http://www.healthdata.org/policy-report/financing-global-health-2015-development-assistance-steady-path-new-global-goals.

Druce, N., and A. Harmer. (2004). *The Determinants of Effectiveness: Partnerships that Deliver Review of the GHP and 'Business' Literature*. London: DFID Health Resource Centre.

Easterly, W. (2006). *The white man's burden: Why the west's effort to aid the rest have done so much ill and so little good*. New York: Penguin.

Ensor, T. (2004). Consumer-led demand side financing in health and education and its relevance for low and middle income countries. *The International Journal of Health Planning and Management*, 19(3), 267–285.

Foundation Center. (2012). *International Grantmaking Update: A snapshot of US foundation trends*. New York: Foundation Center.

Fryatt, R., Mills, A., and Nordstrom, A. (2010). Financing of health systems to achieve the health Millennium Development Goals in low-income countries. *The Lancet*, 375(9712), 419–426.

Gandhi, G. (2015). Charting the evolution of approaches employed by the Global Alliance for Vaccines and Immunizations (GAVI) to address inequities in access to immunization: a systematic qualitative review of GAVI policies, strategies and resource allocation mechanisms through an equity lens (1999–2014). *BMC Public Health*, 15(1), 1–35.

Garrett, L. (2007). Challenge of Global Health. *Foreign Affairs*, 86, 14–38.

Garett, L. (2012). Money or die. A watershed moment for global public health. *Foreign Affairs*. Retrieved from: https://www.foreignaffairs.com/articles/2012-03-06/money-ordie.

GAVI. (2011). Advance market commitment for pneumococcal vaccines: Annual report, 12 June 2009–31 March 2010. Geneva, Switzerland: GAVI. Retrieved from: http://www.gavialliance.org/library/documents/amc/201-pneumococcal-amc-annual-report

GAVI. (2017). Pneumococcal AMC Annual Report: 1 January–31 December 2017. Retrieved from: https://www.gavi.org/library/gavi-documents/amc/2017-pneumococcal-amc-annualreport/.

GAVI. (2020a). GAVI's Business Model. Retrieved from: https://www.gavi.org/our-alliance/operating-model/gavis-business-model.

GAVI. (2020b). Gavi's impact: the vast majority of Gavi-supported countries have introduced the pneumococcal vaccine, reaching more than 183 million children by the end of 2018. Retrieved from: https://www.gavi.org/types-support/vaccine-support/pneumococcal.

GAVI. (2020c). Facts and Figures. Retrieved from: https://www.gavi.org/sites/default/files/document/2020/Gavi-Facts-and-figures-June.pdf.

GAVI. (2020). COVAX Explained. Retrieved from: https://www.gavi.org/vaccineswork/covax-explained.

Gilchrist, S. A., and Nanni, A. (2013). Lessons learned in shaping vaccine markets in low-income countries: a review of the vaccine market segment supported by the GAVI Alliance. *Health Policy and Planning*, 28(8), 838–846.

Graham, J. E. (2019). Ebola vaccine innovation: A case study of pseudoscapes in global health. *Critical Public Health*, 29(4), 401–412.

Grundy, J. (2010). Country-level governance of global health initiatives: An evaluation of immunization coordination mechanisms in five countries of Asia. *Health Policy and Planning*, 25(3), 186–196.

Hafner, T., and Shiffman, J. (2013). The emergence of global attention to health systems strengthening. *Health Policy and Planning*, 28(1), 41–50.

Hanefeld, J. (2008). How have global health initiatives impacted on health equity? *Promotion Education*, 15(1), 19–23.

Hayman, B., and Pagliusi, S. (2020). Emerging vaccine manufacturers are innovating for the next decade. *Vaccine: X*, 5, 100066.

Hill, P. (2011). The Alignment Dialogue: GAVI and its Engagement with National Governments in Health Systems Strengthening. In: *Partnerships and foundations in global health governance* (pp. 76–101). London: Palgrave Macmillan.

Hodge, G. A., and Greve, C. (2007). Public private partnership: An international performance review. *Public Administration Review*, 67(3), 545–558.

Hu, H. W., and Yoshikawa, T. (2017). CEO and board influence on corporate philanthropy in China. In: *Academy of Management Proceedings*. Academy of Management.

Hunter, B. M., and Murray, S. F. (2019). Deconstructing the Financialization of Healthcare. *Development and Change*, 50(5), 1263–1287.

Ifeagwu, S. C., Yang, J. C., Parkes-Ratanshi, R., and Brayne, C. (2021). Health financing for universal health coverage in Sub-Saharan Africa: a systematic review. *Global Health Research and Policy*, 6(1), 1–9.

Ikilezi, G., Augusto, O. J., Dieleman, J. L., Sherr, K., and Lim, S. S. (2020). Effect of donor funding for immunization from Gavi and other development assistance channels on vaccine coverage: Evidence from 120 low and middle income recipient countries. *Vaccine*, 38(3), 588–596.

Institute for Health Metrics and Evaluation. (2018). Financing Global Health 2017. Funding Universal Health Coverage and the Unfinished HIV/AIDS

Agenda. 2018. Retrieved from: http://www.healthdata.org/sites/default/files/files/policy_report/FGH/2018/IHME_FGH_2017_fullreport_online.pdf.

Jamali, D. (2004). Success and failure mechanisms of public private partnerships (PPPs) in developing countries: Insights from the Lebanese context. *The International Journal of Public Sector Management*, 17(5), 414–430.

Johnson, H. L., Deloria-Knoll, M., Levine, O. S., Stoszek, S. K., Hance, L. F., Reithinger, R., Muenz, L. R. and O'Brien, K. L. (2010). Systematic evaluation of serotypes causing invasive pneumococcal disease among children under five: the pneumococcal global serotype project. *PLoS Medicine*, 7(10), e1000348.

Kallenberg, J., Mok, W., Newman, R., Nguyen, A., Ryckman, T., Saxenian, H., and Wilson, P. (2016). Gavi's transition policy: Moving from development assistance to domestic financing of immunization programs. *Health Affairs*, 35(2), 250–258.

Kamya, C., Shearer, J., Asiimwe, G., Carnahan, E., Salisbury, N., Waiswa, P., Brinkerhoff, J., and Hozumi, D. (2017). Evaluating global health partnerships: a case study of a Gavi HPV vaccine application process in Uganda. *International Journal of Health Policy and Management*, 6(6), 327.

Kassim, S., and Abdullah, A. (2017). Pushing the Frontiers of Islamic Finance through Socially Responsible Investment Sukuk. *Al-Shajarah: Journal of the International Institute of Islamic Thought and Civilization (ISTAC)*, 187–213.

Ketkar, S. (2014). Aid securitisation: beyond IFFIm. *International Journal of Public Policy*, 10(1–3), 84–99.

Khanom, N. A. (2010). Conceptual issues in defining public private partnerships (PPPs). *International Review of Business Research Papers*, 6(2), 150–163.

Kickbusch, I., and Quick, J. D. (1998). Partnerships for health in the 21st century. *World health statistics quarterly (Rapport trimestriel de statistiques sanitaires mondiales)*, 51(1), 68–74.

Kieny, M. P., Evans, T. G., Scarpetta, S., Kelley, E. T., Klazinga, N., Forde, I. et al. (2018). Delivering quality health services: A global imperative for universal health coverage. Washington, DC: The World Bank. Retrieved from: https://documents.worldbank.org/en/publication/documents-reports/doc umentdetail/482771530290792652/delivering-quality-health-services-a-glo balimperative-for-universal-health-coverage.

Kim-Farley, R. (1992). Global immunization. *Annual Review of Public Health*, 13(1), 223–237.

Kraak, V. I., Harrigan, P. B., Lawrence, M., Harrison, P. J., Jackson, M. A., and Swinburn, B. (2012). Balancing the benefits and risks of public–private partnerships to address the global double burden of malnutrition. *Public Health Nutrition*, 15(3), 503–517.

Kremer, M. (2000a). Creating Markets for New Vaccines. Part I: Rationale. *Innovation Policy and the Economy*, 1, 35–72.

Kremer, M. (2000b). Creating Markets for New Vaccines. Part II: Design Issues. *Innovation Policy and the Economy*, 1, 73–118.

Kremer, M., and Glennerster, R. (2004). *Strong Medicine: Creating incentives for pharmaceutical research on neglected diseases*. Princeton: Princeton University Press.

Kremer, M., Levin, J., and Snyder, C. M. (2020a). Advance market commitments: Insights from theory and experience. *AEA Papers and Proceedings*, 110, 269–273.

Kremer, M., Levin, J. D., and Snyder, C. M. (2020b). Designing Advance Market Commitments for New Vaccines (No. w28168). National Bureau of Economic Research.

Languille, S. (2017). Public Private partnerships in education and health in the global South: A literature review. *Journal of International and Comparative Social Policy*, 33(2), 142–165.

Lee, K., and Goodman, H. (2002). Global policy networks: The propagation of health care financing reform since the 1980s. In: *Health policy in a globalizing world* (pp. 97–119). Cambridge: Cambridge University Press.

Leitch, S., and Motion, J. (2003). Public-private partnerships: consultation, cooperation and collusion. *Journal of Public Affairs: An International Journal*, 3(3), 273–277.

Light, D. W. (2005). Making practical markets for vaccines. *PLoS Medicine*, 2(10), e271.

Light, D. W. (2009). Advanced Market Commitments: Current realities and alternate approaches. Health Action International (HAI) Europe. Paper Series Reference (03–2009/01). Retrieved from: http://haieurope.org/wp-content/uploads/2010/12/27-Mar-2009-Report-AMC-Current-Realities-Alternate-Approaches.pdf.

Machuca, D. E. (2006). The Pyrrhonist's ἀταραξία and φιλανθρωπία. *Ancient Philosophy*, 26(1), 111–139.

Maimaiti, N., Ahmed, Z., Isa, Z. M., Ghazi, H. F., and Aljunid, S. (2013). Clinical Burden of Invasive Pneumococcal Disease in Selected Developing Countries. *Value in Health Regional Issues*, 2(2), 259-263.

Malhame, M., Baker, E., Gandhi, G., Jones, A., Kalpaxis, P., Iqbal, R., Momeni, Y., and Nguyen, A. (2019). Shaping markets to benefit global health–A 15-year history and lessons learned from the pentavalent vaccine market. *Vaccine: X*, 2, 100033.

Malmborg, R., Mann, G., Thomson, R., and Squire, S. B. (2006). Can public-private collaboration promote tuberculosis case detection among the poor and vulnerable? *Bulletin of the World Health Organization*, 84, 752–758.

Marchal, B., Cavalli, A., and Kegels, G. (2009). Global health actors claim to support health system strengthening—is this reality or rhetoric? *PLoS Medicine*, 6(4), e1000059.

McCoy, D. (2009). Global health initiatives and country health systems. *The Lancet*, 374, 1237–1237.

McCoy, D., Kembhavi, G., Patel, J., and Luintel, A. (2009a). The Bill & Melinda Gates Foundation's grant-making programme for global health. *The Lancet*, 373(9675), 1645–1653.

McCoy, D., Chand, S., and Sridhar, D. (2009b). Global health funding: How much, where it comes from and where it goes. *Health Policy and Planning*, 24, 407–417.

McCoy, D., and Brikci, N. (2010). Taskforce on innovative international financing for health systems: What next? *Bulletin of the World Health Organization*, 88, 478–480.

McGoey, L., Reiss, J., and Wahlberg, A. (2011). Editors' Introduction: The global health complex. *BioSocieties*, 6, 1–9.

McKinsey and Company. (2005). Building Effective Public Private Partnerships: Lessons Learned from the Jordan Education Initiative. Retrieved from: http://www.weforum.org/pdf/JEI/JEIreport.pdf.

Miller, J. B. (2000), *Principles of Public and Private Infrastructure Delivery*. London: Kluwer Academic Publishers.

Mills, A. (1983). Vertical vs horizontal health programmes in Africa: Idealism, pragmatism, resources and efficiency. *Social Science and Medicine*, 17, 1971–1981.

Mills, A. (2005). Mass campaigns versus general health services: What have we learnt in 40 years about vertical versus horizontal approaches? *Bulletin of the World Health Organization*, 83, 315–316.

Milstien, J. B., Batson, A., and Wertheimer, A. I. (2005). Vaccines and Drugs: Characteristics of Their Use to Meet Public Health Goals. Health, Nutrition and Population Discussion Papers, The World Bank. Retrieved from: https://openknowledge.worldbank.org/handle/10986/13696.

Miraftab, F. (2004). Public-private Partnerships: The Trojan Horse of Neoliberal Development? *Journal of Planning Education and Research*, 24(1), 89–101.

Mitchell-Weaver, C., and Manning, B. (1991). Public-private partnerships in third world development: A conceptual overiew. *Studies in Comparative International Development*, 26(4), 45–67.

Moran, M. (2008). 'The 800 pound gorilla': The Bill & Melinda Gates Foundation, the GAVI Alliance and philanthropy in international public policy. In: *Annual Meeting of the International Studies Association*. San Francisco.

Moran, M. (2011). Private foundations and global health partnerships: Philanthropists and 'partnership brokerage'. In: *Partnerships and foundations in global health governance* (pp. 123–142). London: Palgrave Macmillan.

Moran, M. (2014). *Private foundations and development partnerships: American philanthropy and global development agendas.* London and New York: Routledge.

Moran, M., and Stone, D. (2016). The new philanthropy: Private power in international development policy? In: *The Palgrave Handbook of International Development* (pp. 297–313). London: Palgrave Macmillan.

Muraskin, W. (2002). The last years of the CVI and the Birth of the GAVI. In: *Public-private partnerships for public health* (pp. 115–168). Cambridge, MA: Harvard University Press.

Muraskin, W. (2005). *Crusade to immunize the world's children.* Los Angeles: USC Marshall Global BioBusiness Initiative.

Naimoli, J. F. (2009). Global health partnerships in practice: Taking stock of the GAVI Alliance's new investment in health systems strengthening. *International Journal of Health Planning and Management,* 24(1), 3–25.

Nijkamp, P., Van Der Burch, M., and Vindigni, G. (2002). A comparative institutional evaluation of public-private partnerships in Dutch urban land-use and revitalisation projects. *Urban Studies,* 39(10), 1865–1880.

O'Brien, K. (2017). Current Status of PCV Use and WHO Recommendations. Retrieved from: https://www.who.int/immunization/sage/meetings/2017/october/01_17_October_2017_Presentation_01_OBrien_SAGE_PCV.pdf.

Oliveira-Cruz, V. (2008). Financing primary health care. *id21 insights health,* 12, 1–2.

Ooms, G., Van Damme, W., and Temmerman, M. (2007). Medicines without Doctors: Why the Global Fund Must Fund Salaries of Health Workers to Expand AIDS Treatment. *PLoS Medicine,* 4(4), 605–608.

Osborne, S. (2000). *Public-private partnerships: Theory and practice in international perspective.* London: Routledge.

Pongsiri, N. (2002). Regulation and public-private partnerships. *International Journal of Public Sector Management,* 15(6), 487–495.

Ravishankar, N., Gubbins, P., Cooley, R. J., Leach-Kemon, K., Michaud, C. M., Jamison, D. T., and Murray, C. J. (2009). Financing of global health: Tracking development assistance for health from 1990 to 2007. *The Lancet,* 373, 2113–2124.

Reddy, C. L., Peters, A. W., Jumbam, D. T., Caddell, L., Alkire, B. C., Meara, J. G., and Atun, R. (2020). Innovative financing to fund surgical systems and expand surgical care in low-income and middle-income countries. *BMJ Global Health,* 5(6), e002375.

Reich, M. R. (Ed.). (2002). *Public-private partnerships for public health.* Cambridge, MA: Harvard Center for Population and Development Studies.

Richter, J. (2004). Public–private partnerships for Health: A trend with no alternatives? *Development,* 47, 43–48.

Roth, G. A., Abate, D., Abate, K. H., Abay, S. M., Abbafati, C., Abbasi, N., Borschmann, R. et al. (2018). Global, regional, and national age-sex-specific mortality for 282 causes of death in 195 countries and territories, 1980–2017: A systematic analysis for the Global Burden of Disease Study 2017. *The Lancet*, 392(10159), 1736–1788.

Ruckert, A., and Labonté, R. (2014). Public–private partnerships (PPPs) in global health: The good, the bad and the ugly. *Third World Quarterly*, 35(9), 1598–1614.

Rushton, S., and Williams, O. D. (Eds.). (2011). *Partnerships and foundations in global health governance*. London: Palgrave Macmillan.

Ryman, T. K., Dietz, V., and Cairns, K. L. (2008). Too little but not too late: results of a literature review to improve routine immunization programs in developing countries. *BMC Health Services Research*, 8(1), 1–11.

Samb, B., Evans, T., Dybul, M., Atun, R., Moatti, J. P., and Nishtar, S. (2009). World Health Organization maximising positive synergies collaborative group: An assessment of interactions between global health initiatives and country health systems. *The Lancet*, 373(9681), 2137–2169.

Santos Rutschman, A. (2020). The COVID-19 vaccine race: Intellectual property, collaboration (s), nationalism and misinformation. *Washington University Journal of Law and Policy*, 64.

Savas, E. S. (2000). *Privatization and Public Private Partnerships*. New York: Seven Bridges Press.

Scharle, P. (2002). Public-private partnership (PPP) as a social game. *Innovation: The European Journal of Social Science Research*, 15(3), 227–252.

Schwalbe, N., and El-Ziq, I. (2010). GAVI's advance market commitment. *The Lancet*, 375(9715), 638–639.

Snyder, C. M., Begor, W., and Berndt, E. R. (2011). Economic perspectives on the advance market commitment for pneumococcal vaccines. *Health Affairs*, 30(8), 1508–1517.

Spackman, M. (2002). Public–private partnerships: lessons from the British approach. *Economic Systems*, 26(3), 283–301.

Sridhar, D., and Tamashiro, T. (2009). Vertical funds in the health sector: Lessons for education from the Global Fund and GAVI. Papers commissioned for the EFA Global Monitoring Report.

Sridhar, D., and Woods, N. (2013). Trojan multilateralism: global cooperation in health. *Global Policy*, 4(4), 325–335.

Stein, F., and Sridhar, D. (2018). The financialisation of global health. *Welcome Open Research*, 3(17).

Stone, D. (2013). *Knowledge actors and transnational governance: The Public-Private Policy Nexus in the Global Agora*. Palgrave Macmillan.

Storeng, K. T. (2014). The GAVI Alliance and the 'Gates approach' to health system strengthening. *Global Public Health*, 9(8), 865–879.

Sulek, M. (2010). On the modern meaning of philanthropy. *Nonprofit and Voluntary Sector Quarterly*, 39(2), 193–212.

Suwantika, A. A., Zakiyah, N., Kusuma, A. S., Abdulah, R., and Postma, M. J. (2020). Impact of switch options on the economics of pneumococcal conjugate vaccine (PCV) introduction in Indonesia. *Vaccines*, 8(2), 233.

Szlezak, N. K., Bloom, R. B. R., Jamison, D. T., Keusch, G. T., Michaud, C. M., Moon, S., and Clark, W. C. (2010). The global health system: Actors, norms, and expectations in transition. *PLoS Medicine*, 7(1), e1000183.

Tasslimi, A., Nakamura, M. M., Levine, O., Knoll, M. D., Russell, L. B., and Sinha, A. (2011). Cost effectiveness of child pneumococcal conjugate vaccination in GAVI-eligible countries. *International Health*, 3(4), 259–269.

U.S. Health and Human Services. (2020). U.S. Government Engages Pfizer to Produce Millions of Doses of COVID-19 Vaccine. Retrieved from: https://www.hhs.gov/about/news/2020/07/22/us-government-eng ages-pfizer-produce-millions-doses-covid-19-vaccine.html.

United Nations Children's Fund (UNICEF). (2018). Pneumococcal Conjugate Vaccine: Supply and Demand Update. UNICEF Supply Division. Retrieved from: https://www.unicef.org/supply/reports/pneumococcal-con jugate-vaccine-pcv-market-update.

United Nations Children's Fund (UNICEF). (2019). *Levels and trends in child mortality, report 2019: for Child Mortality Estimation: Estimates developed by the UN Inter-Agency Group for Child Mortality Estimation*. New York: UNICEF. Retrieved from: https://data.unicef.org/resources/levels-and-tre nds-in-child-mortality/.

Wahl. B., O'Brien, K. L., Greenbaum, A., Majumder, A., Liu, L., Chu, Y., Lukšić, I., Nair, H., McAllistar, D.A., Campbell, H., et al. (2018). Burden of Streptococcus pneumoniae and Haemophilus influenzae type b disease in children in the era of conjugate vaccines: Global, regional, and national estimates for 2000–15. *The Lancet Global Health*, 6(7), e744–e757.

Widdus, R. (2001). Public-private partnerships for health: Their main targets, their diversity, and their future directions. *Bulletin of the World Health Organization*, 79, 713–720.

Williams, A. T. (1997). Regional Governance: Contemporary Public Private Partnerships in the South (Doctoral dissertation, Virginia Commonwealth University).

World Health Organisation. (2011). Panacea Biotec DTP-based combination and monovalent hepatitis B vaccines delisted from WHO list of prequalified vaccines. Retrieved from: www.who.int/immunization_standards/vaccine_q uality/DTP_mono_hepb_aug2011/en/.

World Health Organization (WHO). (2013). Global vaccine action plan 2011–2020. Retrieved from: http://www.path.org/publications/files/OTP_dov_ gvap_2011_20.pdf.

World Health Organization (WHO). (2017a). Tracking universal health coverage: 2017 Global Monitoring Report. World Health Organization.

World Health Organization (WHO). (2017b). The Power of Vaccines: Still Not Fully Utilized. Retrieved from: https://www.who.int/publications/10-year-review/vaccines/en/.

World Health Organization (WHO). (2019). Pneumococcal conjugate vaccines in infants and children under 5 years of age: WHO position paper—February 2019. Weekly Epidemiological Record 2019, 94, 85–103. Retrieved from: https://apps.who.int/iris/handle/10665/310970.

World Health Organization (WHO). (2020a). Immunization Coverage. Retrieved from: https://www.who.int/en/news-room/fact-sheets/detail/immunization-coverage.

World Health Organization (WHO). (2020b). List of Prequalified Vaccines. Geneva, Switzerland: World Health Organization. Retrieved from: https://extranet.who.int/pqweb/vaccines/list-prequalified-vaccines.

Youde, J. (2013). The Rockefeller and Gates Foundations in global health governance. Global Society, 27(2), 139–158.

Zammit, A. (2003). Development at risk: Rethinking UN-business partnerships. Geneva: UNRISD and South Centre.

Zerhouni, E. (2019). GAVI, the vaccine alliance. Cell, 179(1), 13–17.

CHAPTER 4

Performance-Based Financing

INTRODUCTION

The current humanitarian landscape is marked by the growing role of private sector funding. It is often said that these sources of funding are often justified as a necessary strategy to fill the growing humanitarian financing deficit which is considered to be beyond the capability of traditional donor financing. The desire to crowd in alternative funding sources and engage private sector partners is centered on the notion of "harnessing the power of business" where the role of the private sector in humanitarian aid is seen to be underutilized also, there is an increased expectation for the private sector to devote more funds to support humanitarian initiatives (High-Level Panel on Humanitarian Financing 2016).

A type of financial mechanism that is expected to support and facilitate more private sector funding in the humanitarian sector is performance-based financing (PBF). According to Bertone et al. (2018), one of the aims of PBF is to improve healthcare service provision by providing bonuses to service providers based on audited quantity of outputs produced and modified by quality indicators. These mechanisms have been increasingly implemented in low and middle-income countries, as well as being used in fragile and humanitarian settings. According to the

© The Author(s), under exclusive license to Springer Nature Switzerland AG 2021
M. Ahmed, *Innovative Humanitarian Financing*,
Palgrave Studies in Impact Finance,
https://doi.org/10.1007/978-3-030-83209-4_4

literature on PBF, it is unlikely to be a homogeneous intervention; moreover, its impacts and modalities will be highly context dependent (Witter et al. 2012).

There is a dearth of literature on how PBF has been implemented, adapted to different contexts and how that may influence the adoption, adaptation, design, application as well as impacts of PBF programs (Renmans et al. 2017). With a case study of a PBF program designed for fragile contexts, this chapter will examine how this type of mechanism emerged, how it has been adapted to humanitarian settings, what the opportunities and challenges are and what lessons can be learnt.

With this objective in mind, this chapter is organized as follows: the next section will provide a literature review on environmental, social and governance in addition, the section will also analyze the literature on impact investing, outcomes-based performance management and impact bonds. Section three will evaluate the world's first humanitarian impact bond using the case study method to assess how this innovative type of financial instrument will be used for humanitarian purposes in Africa. The section will also give a brief background on the type of health service provision the bond intends to raise funds for, which in this case is to address some of the healthcare challenges faced by disabled persons' in fragile contexts. Also, the section will provide an in-depth analysis of the case. Section four will conclude.

Literature Review

Environmental, Social and Governance (ESG)

The concept of environmental, social and governance (ESG) is generally used in corporate settings to represent a set of relevant ESG-related factors that take into consideration the assessment of the long-term sustainability of investments through the integration of traditional economic and financial parameters (Taliento et al. 2019). ESG elements aim to measure additional dimensions of corporate performance which are not revealed in the accounting data (Bassen and Kovács 2008). ESG therefore captures a more holistic element of non-financial data that can be utilized to evaluate factors such as the overall capabilities of the management of a company as well as support risk management (Galbreath 2013).

In essence, ESG is rooted in the stakeholder theory. Stakeholder theory stipulates that any party including employees, customers, shareholders, investors and society at large should be considered as a stakeholder of an organization furthermore, stakeholders are parties that have an impact or are being affected by an organization (Carroll 1991; Freeman 1984; Wood 1991) nonetheless ESG factors are important to stakeholders (Sila and Cek 2017). According to the United Nations Principles for Responsible Investment, ESG elements refer to three different but related fields within the "social awareness" sphere.

The first factor of ESG relates to environmental issues such as climate change, greenhouse gas emissions, pollution, deforestation, waste and the exploitation of resources (Tamimi and Sebastianelli 2017). In recent times, both internal and external stakeholders are increasingly showing interest in the environmental performance of organizations due to the impact of pollution being created by them. For example, internal stakeholders such as employees might be impacted by pollution in their working environments while external stakeholders such as environmental activist groups, government regulators, shareholders, investors, customers, suppliers, the local community and others will also take an interest in corporate pollution (Jasch 2006).

The second factor of ESG relates to the social conditions within a working environment including health and safety, relationships with employees and diversity (Tamimi and Sebastianelli 2017). Corporate social practices are a firm's structure of social responsibility principles, procedures of social responsiveness and policies, programs and tangible outcomes relating to the firm's social relationships (Wood 1991). Furthermore, it can also be defined as a construct that highlights a company's responsibility to multiple stakeholders such as employees, shareholders and the wider community as a whole, it has also been found that organizations with high social performance are able to better attract eligible employees (Turban and Greening 1997).

The third factor of ESG relates to corporate governance practices including managerial remuneration, the composition of the Board of Directors, audit procedures and the behavior of Senior and Corporate Executives in terms of compliance with the law in addition to the ethical principles and codes of conduct (Tamimi and Sebastianelli 2017). A good corporate governance system is an essential factor to optimize the performance of a business in the best interest of shareholders, limit agency costs and support the future existence of corporations (Fama and Jensen 1983).

Furthermore, corporate governance is not only concerned with the goal of providing long-term shareholder value but also takes into consideration the interests of other stakeholders (Tarmuji et al. 2016).

Impact Investing: An Overview

The term "impact investing" was coined in 2007 at a gathering of finance, development and philanthropy leaders organized by the Rockefeller Foundation in Italy (Hajri and Jackson 2012). Although the definition of impact investing can be rather contentious, it can be broadly defined as investments that have a dual purpose of generating both financial returns and delivering social and environmental impact (Louche et al. 2012). The main rationale for impact investments is achieving financial returns should not be the sole objective in business decisions and investment criteria.

Although impact investing is often compared to or thought of as a form of socially responsible investing (SRI), the main difference is, SRI is a technique that screens investments for environmental, social and governance factors whereas impact investing seeks to yield positive and measurable impacts as well as financial returns (Geobey et al. 2012; Weber 2016). Impact investors go beyond conventional SRI as they aim to proactively allocate capital to businesses or projects to enhance environmental or social objectives (Nicholls 2010a, b).

Impact investing is defined as "the active investment of capital in businesses and funds that generate positive social and/or environmental impacts, as well as financial returns (from principal to above market rate) to the investor" by the Canadian Task Force on Social Finance (2010). Furthermore, Drexler et al. (2013) describe impact investing as "an investment approach that intentionally seeks to create both financial return and positive social or environmental impact that is actively measured." According to La Torre and Calderini (2018), impact investing can be thought of as an investment approach rather than a separate asset class and secondly, in order for an investment to be classified as an "impact investment" outcomes need to be measured.

Essentially impact investing can be seen as a tool to unlock capital to intentionally mobilize it toward firms and projects that generate social and environmental benefits while simultaneously generating a financial return for investors (Hajri and Jackson 2012). As such, impact investing is the confluence of philanthropic objectives and mainstream financial decision-making (Hochstadter and Scheck 2015). However, what distinguishes impact investing from traditional grant funding and philanthropy is the objective to yield a financial return also, what differentiates impact investing from conventional investments is the focus on non-financial impact (Addis et al. 2013; Wong 2012).

Although a relatively newly defined term, the concept of intentionally deploying capital to produce both financial and non-financial returns as well as aiming to achieve social outcomes through investments is not a novel concept (Nicholls 2010a; O'Donohoe et al. 2010; Saltuk 2011). However, over the past two decades, the topic has become more relevant on the international agenda (Wong and Yap 2019).

According to Bugg-Levine and Emerson (2011), it can be said that impact investing can trace its origins to the centuries-old tradition of the wealthy being held responsible for the welfare of wider society. Furthermore, the ideals of impact investing can be traced to the Quakers in seventeenth-century England who aimed to align both their investment and consumption decisions with their values (Bugg-Levine and Emerson 2011). Also, multilateral examples include development finance institutions such as the UK's Commonwealth Development Corporation established in 1948 and the World Bank's International Finance Corporation established in 1956 (O'Donohoe et al. 2010).

In recent times, the notion that investors can pursue financial returns while "doing good" both socially and environmentally has garnered interest among industry practitioners and policymakers alike (Bugg-Levine and Emerson 2011; Social Impact Investment Taskforce 2014; Trelstad 2009).

More recently, there have been concerted efforts to build an impact investing market (O'Donohoe et al. 2010; Saltuk 2011) and in the past couple of years, a convergence of factors has fueled the impact investing market. Factors such as the global financial crisis, the desire to utilize philanthropic capital and public money to address social and environmental issues, the track record of business models generating sustainable and scalable returns and the wealth transfer to new generations with a strong interest in impact investing (Hajri and Jackson 2012).

The renewed interest in the latest iteration of impact investing in combining social and environmental impacts with financial return has been part of a growing trend since the 2007/2008 global financial crisis. Post-crisis questions have been raised about the societal benefit of financial markets and how they operate (Shiller 2013; Zingales 2015). More specifically, there is a growing need to design and develop new investment opportunities to create blended returns and wider societal shared value (Porter and Kramer 2011; Lehner 2016; Weber and Feltmate 2016; Jacobs and Mazzucato 2016).

Impact investment discourse has primarily been driven by practitioners. However, academic studies have highlighted the stark disparities in the conceptualization of impact investing such as the double and triple bottom line, mission-related investing, program-related investment, blended-value and economically targeted investing (La Torre an Calderini 2018). The variations and inconsistencies in definitional and terminological aspects have resulted in a major impediment to the growth of the impact investing industry as there is a lack of a common understanding (GIIN 2017). Furthermore, without conceptual clarity, it is difficult for the impact investing industry to not only gain legitimacy but also difficult for theories to be developed—further hindering the growth of the industry.

As noted by Moore et al. (2012) and Nicholls (2010a), academic research on impact investment is at the early stages of developing. Several attempts in recent years have been made by academic researchers to contribute to this emergent field. For example, attempts by Hochstadter and Scheck (2015) helped to clarify the concept of impact investing by investigating works by a number of academics and practitioners to highlight both the similarities and inconsistencies in definitional, terminological and strategic levels. Rizzello et al. (2016) illustrate the academic landscape of impact investing by mapping contributions, areas of inquiries and future research agendas through the exclusive analysis of peer-reviewed work. Care and Wendt (2018) shed light on impact investing by summarizing the literature and provide an assessment of financial instruments and investment opportunities available in the market in addition to the actors involved.

Outcomes-Based Performance Management (OBPM)

Outcomes-based performance management (OBPM) began in the 1990s (Schalock 2001; Perrin 1998), but in the past few decades, various forms have emerged. According to Lowe and Wilson (2017), OBPM has become the basis of payment by results. More recently, OBPM has become the basis of social impact bonds in which people or organizations are rewarded by the "outcomes" that they deliver (Cabinet Office 2015; National Audit Office 2015).

A "payment-by-results" or "pay-for-success" contract can be defined as a contract whereby the government agrees to pay for improved social outcomes if the project is deemed successful or meets the metrics agreed upon beforehand. The term "payment by results" (PbR) is a description of an OBPM whereby payment is dependent (partially or entirely) on the achievement of specified goals or targets (Albertson and Fox 2018) and this is derived from the OBPM methodology that uses outcomes as a tool to evaluate the performance and effectiveness of social policy interventions (Lowe 2013).

Proponents of using the OBPM method argue its main appeal is to "focus the attention of those delivering social policy interventions on those whom they serve" (Lowe and Wilson 2017). OBPM suggests that the effectiveness of social policy interventions should be evaluated through the lens of impact being made on the lives of the beneficiaries for whom they were designed for. Furthermore, it suggests that the service providers of those interventions ought to have their performance compensated or punished depending on whether the desired outcomes have been achieved (Bovaird 2014; Brien 2011).

When evaluating the effectiveness of OBPM, research has tended to fall into one of two camps (Lowe and Wilson 2017). On the one hand, quantitative-based research that conducted large-scale econometric analysis of the impact of performance data has tended to provide supporting evidence of the effectiveness of OBPM. On the other hand, qualitative-based research, based on interviews with those who deliver social interventions, tends to find a lack of the effectiveness of OBPM. In fact, researchers have found that not only is OBPM detrimental do the quality of services delivered, it also distorts the practice of social interventions and evidence of "gaming" was found (Lowe and Wilson 2017).

Lowe and Wilson (2017) have found that the theoretical underpinnings of OBPM are flawed in itself as there is a reliance on the simplification of outcomes in order to fit the model. According to the researchers, there are two fundamental issues with this model. This linear type thinking does not take into consideration the complexities of life and the lived experiences of the benefactors of social interventions thereby turning it into a "game," using simple measurable outcomes (often linear), does not accurately capture the genuine impact of programs, instead outcomes are simplified in order to be measurable without accounting for the complex system in which they operate in (Lowe and Wilson 2017). This in turn encourages those who deliver social interventions to collect and produce data that demonstrates performance in order to be rewarded.

Impact Bonds

Impact bonds, a well-known type of impact investing tool with many variations, are public–private partnerships that aims to improve social outcomes for service users/beneficiaries. They are outcome-based contracts involving many parties whereby investors provide upfront capital to cover the cost for a provider to deliver a particular service. Impact bonds create a contract between public and private stakeholders where in most cases, public entities identify quantifiable social outcomes they would like to be achieved.

In a typical impact bond, a contract is entered into by a public institution, a social service providing organization and private investors—to deliver individualized support for a vulnerable and marginalized group (Andreu 2018). Beneficiaries of such support can be groups such as former offenders, homeless persons, chronically ill persons, marginalized indigenous persons or youth who are at-risk of early deprivation, poverty and abuse (Dear et al. 2016). According to Andreu (2018), one of the advantages of impact bonds is, they are said to help the actors involved better understand how and to what extent the interventions are effective and which types of interventions public institutions should be providing and scaling up. Therefore, impact bonds are believed to improve accountability and the effective use of taxpayers' money.

In simple terms, impact bonds bring together five main stakeholders: (i) outcome payers, (ii) service providers, (iii) investors, (iv) intermediaries and (v) evaluators (Government Outcomes Lab 2018a):

i. Outcome payers will firstly stipulate a specific social issue they intend to tackle. They will then set out a criterion of certain outcomes that they would like to be achieved in a program and express a "willingness to pay" for them (Government Outcomes Lab 2018a).

ii. Service providers are selected to offer a service or intervention to achieve the outcomes set out by the outcome payers. As impact bonds are outcomes-based contracts, the payment to service providers is contingent on whether the outcomes are achieved.

iii. Investors provide upfront capital for a service provider to deliver a particular social service over the duration of the project. Investors bear the financial risk as repayment to investors (wholly or partly) depends on whether the pre-agreed outcomes are achieved. In return, investors are offered the opportunity to earn a financial return if the intervention succeeds when social outcomes are met; however, if an intervention fails, investors may lose some or all of their principal investments and make little or no profit (Griffiths and Meinicke 2014). Social and financial returns are therefore aligned as investors receive higher financial returns if there are greater improvements in social outcomes (Nicholls and Tomkinson 2013). For impact bonds, there are usually more than one type of investor involved such as foundations, banks, private investors, unlike other outcomes-based contracts (Government Outcomes Lab 2018a).

iv. Impact bonds sometimes use intermediaries such as consultants, social investment fund managers, performance management experts, special purpose management companies, to provide specific services (Government Outcomes Lab 2018a). Intermediaries are usually the connecting force that link investors, government entities, not-for-profit service providers and evaluators together (Kasper and Marcoux 2014).

v. Independent evaluators are selected to determine whether a project or intervention has achieved the social objectives based on the criteria set out by the contractors, which will be decided in the early stages of a project (Government Outcomes Lab 2018a).

There are three main types of impact bonds: social impact bonds, development impact bonds and humanitarian impact bonds. What ultimately differentiates the various types of impact bonds is who the outcome funders are as well as the context in which the bonds are implemented which will be discussed further below.

Social Impact Bonds (SIBs)

A form of impact investing and type of impact bond, Social Impact Bonds (SIBs) are an emergent investment mechanism designed to appeal to impact investors (Ragin and Palandjian 2013). SIBs differ from standard debt contracts because they do not offer a fixed rate of return (or a return which is linked to some other interest rates). They are instead, equity-like instruments in which returns depend on the outcome (in this case the social impact) of a certain investment project. According to So and Jagelewski (2013) an SIB is a pay-for-success contractual obligation with the aim of addressing the root causes (not the treatment) of social issues so that savings can be made for the Government in addressing them. SIBs can therefore be considered as a way of "conducting social policies and spending public money" and part of the wider phenomenon of impact investing (Chiapello 2015).

The initial idea of SIBs was conceived by UK think-tank Council on Social Action (CoSA) in 2007 and championed by former Prime Minister Gordon Brown (Council on Social Action 2008). SIBs were first pioneered and used in the United Kingdom in 2010 (Humphries 2013) to tackle the issue of recidivism through an intervention program that was piloted in Peterborough prison. The aim of the program was to reduce the reoffending rate among male prisoners of a targeted population group. Since then, SIBs have also been used to tackle various social issues such as Early Childhood Development, Unemployment, Foster Care Avoidance, Healthcare, Homelessness, Youth Unemployment and Education and Youth Homelessness.

SIBs are characterized by: (i) the participation of both the private and public sector in public private partnerships; (ii) an initial monetary investment; and (iii) an action program (Trotta et al. 2015). SIB investors expect both financial and social returns, similar to impact investors (O'Donohoe et al. 2010) also, a key feature of SIBs is the prominence placed on measurement and evaluation through outcome metrics (Rangan and Chase 2015).

It has been argued that SIBs represent an innovative market-based funding model that in a transparent way leverages traditional capital to impact investors (Ragin and Palandjian 2013). SIBs have also presented an opportunity for private investment to fund the delivery of social services (Fox and Albertson 2011). In general, SIBs are representative of a financing model that is adaptable to various social issues (Fox and Albertson 2011; Hedderman 2013; Fitzgerald 2013; Stoesz 2014). Moreover, according to Schinckus (2015), "by redesigning social programs through market-based solutions, SIBs enhance transparency and evaluation of expenditures made by government, and they can stabilize economic activity and they can contribute to the self-realization of disadvantaged people." In the past couple of years, the pay-for-success model utilized by SIBs has inspired a variety of financial impact-oriented financial structures (La Torre and Calderini 2018).

Although SIBs have been hailed as innovative financing mechanisms that are used to fund social issues, some criticisms have been raised. For instance, the complexity of SIB models, the significant costs incurred in designing and structuring them and whether they offer the social outcomes intended for the end users/beneficiaries have been pointed out by various critics (Arena et al. 2016; Berndt and Wirth 2018; Giacomantonio 2017).

There have been over one-hundred SIBs contracted in a social policy and health context largely in Anglo-Saxon countries (Andreu 2018) and SIBs have been used globally in countries such as Australia, Belgium, Canada, India, Israel, Ireland, the Netherlands, Portugal and the United States of America (Finance for good 2015). Although some have stated in the past that SIBs are the most popular approach to finance initiatives in the social sector (Spiess-Knafl and Scheck 2017), the global market for SIBs remains relatively small at approximately USD$514 million in contract size and USD$210 million in investment (Floyd 2017).

Development Impact Bonds

Development Impact Bonds (DIBs) are PbR instruments that are adapted and modeled on SIBs to finance public services in developing countries (Clarke et al. 2015). DIBs are primarily used in low and middle-income countries and what distinguishes them from SIBs is the payer of the outcomes. For DIBs, the outcome payers are either foreign governments, multilateral aid agencies or philanthropic institutions (Government Outcomes Lab, 2018a, b). They were first conceptualized in 2012 but launched in 2015 with the first two DIBs being launched in Peru and India to support agricultural development and the education of girls' respectively (Government Outcomes Lab 2018a, b).

According to Wilson (2014), by introducing private sector players, DIBs aim to improve the efficacy of conventional donor-funded projects by "shifting the focus on to implementation quality and the delivery of successful results" as the private sector are thought to be more able to take on risks associated with innovation than the public sector. Moreover,

DIBs as financial instruments help facilitate new funding sources from private investors to help improve development outcomes (Wilson 2014).

DIBs involve four main parties: (i) investors who provide the start-up capital for a program or intervention; (ii) service providers who utilize the capital provided from investors to implement programs; (iii) outcome funders/outcome payers who repay investors their principal payment and interest if results (previously agreed upon) are reached; and (iv) an independent third party who evaluate the results and program prior to the outcome funders repaying investors if the program is deemed a success and achieved its objectives (Oroxom et al. 2018).

Humanitarian Impact Bonds

Although a relatively new instrument, impact bonds in recent times have expanded into the humanitarian sector where so-called Humanitarian Impact Bonds (HIBs) have been introduced to achieve both positive social outcomes as well as financial but in a humanitarian context. It has been argued that the uptake, effectiveness and macroeconomic significance of impact bonds marks a shift in the broader discourse of humanitarian reasoning, moreover the underlying context of this shift is an awareness of unfettered market-based solutions to humanitarian aid and development (Mitchell 2017). The next section will further assess the world's first HIB and will discuss whether humanitarian principles and economic value can both be achieved simultaneously.

CASE STUDY: INTERNATIONAL COMMITTEE OF THE RED CROSS (ICRC) HUMANITARIAN IMPACT BOND

Healthcare Challenges for Disabled Persons' in Fragile Contexts

Access to appropriate healthcare is an important human right enshrined in the United Nations Convention of the Rights of Persons with Disabilities (Stein et al. 2009), the convention underlines the rights of individuals with functional problems to adequate rehabilitation services to ensure their ability to fully participate in society (Eide 2013). Yet for disabled persons living in regions that are resource-constrained, unmet health needs not only negatively impacts their quality of life but also threatens their basic survival (Mirza 2015). Within this context, access to healthcare for disabled persons' is an essential concern (Tomlinson et al. 2009).

Gaps in adequate healthcare services particularly impacts disabled persons belonging to marginalized groups such as forced migrants. In humanitarian camps, disabled persons are frequently not addressed during health sector planning and encounter significant barriers to accessing basic health services and overall, the availability of specialized health and rehabilitation services is lacking (Shivji 2010). The majority of health-related research in humanitarian camps has sidelined disability issues and instead focused on other health issues such as infectious and communicable diseases (Roca et al. 2011), reproductive health (Howard et al. 2011), dietary interventions (Khatib et al. 2010) and mental health conditions (Mollica et al. 1993). Consequently, traditional health services in humanitarian camps have concentrated on preventative and curative responses to acute conditions in addition, population-based primary care and preventative interventions have seldom taken into account the needs of disabled persons (Mirza 2015). Although acute infectious diseases have always been of concern, chronic non-infectious conditions and long-term impairments are increasing due to a longer life expectancies among displaced populations also, as a result of a significant number of displaced persons now originating from middle-income countries (Spiegel et al. 2010). A study conducted by Mirza (2015) on healthcare access for forced migrants with disabilities within humanitarian camps highlighted several issues and barriers for disability-specific health needs such as misperceptions about the health-related needs of disabled persons, the specialized health needs of disabled persons' falling outside of the "social minimum" of humanitarian healthcare and concerns about the distributional ethics in regard to disability-inclusive healthcare.

In predominantly western disability discourse, the difficult realities of the lives of disabled persons in impoverished regions of the world are not addressed adequately (Meekosha 2011). For example, in the case of specialized health and rehabilitation care, there tends to be concerns about over medicalization, that may seem reasonable in a context where there is universal access to basic healthcare and abundancy of medical specialists but is misplaced in contexts where trained healthcare professionals are limited (Mirza 2015). For instance, at the local level, only one-third of low-income countries are reported to have access to rehabilitation services provided by trained professionals (WHO 2010) and according to the World Disability Report, in low and middle-income countries, access to quality rehabilitation services is usually low (WHO 2011). Comprehensive efforts are therefore needed in the humanitarian sector in order to develop financial instruments targeting positive health and social care outcomes for disabled persons living in developing or fragile contexts.

Case Overview

In 2017, the International Committee of the Red Cross (ICRC) launched the world's first humanitarian impact bond (HIB) for the provision of physical rehabilitation services in post-conflict African countries (Carè and Ferraro 2019). Proceeds raised from the 5-year HIB were used to build and operate new physical rehabilitation centers in the Democratic Republic of Congo (Kinshasa), Mali (Mopti) and Nigeria (Maiduguri) (Andreu 2018). The outcomes and achievements of the program will then be independently evaluated and verified by a third party, in this case, Philanthropy Associates (Clarke et al. 2018).

As the world's largest provider of physical rehabilitation services in developing and fragile states, the ICRC sought to expand its Physical Rehabilitation Program through the funds raised from impact bonds. The program, which has provided physiotherapy and access to mobility devices since 1979, operates in over 100 centers worldwide (International Committee of the Red Cross 2016a, b). The aim of the HIB was to fund the support of at least 3600 persons who have physical disabilities caused by war, natural disasters, congenital impairments or disabling diseases and help them regain mobility (Government Outcomes Lab 2018b). Proceeds raised from the bond were used for the construction of rehabilitation centers and the support of the operation of the centers, covering costs such as training additional staff and implementing new Information Technology tools for two years after the opening of the centers (Andreu 2018).

The five outcome funders/donors for the bond were the government agencies of Belgium, Switzerland, Italy and the United Kingdom and the Spanish headquartered La Caixa Foundation (Clarke et al. 2018) also, the government of the Netherlands provided grant financing for the HIB's design and structuring (Government Outcomes Lab 2018b). In general, the various donors were targeting different outcomes, for example, La Caixa Foundation paid for the construction of new rehabilitation centers while the rest of the government outcome donors funded the rehabilitation of the beneficiaries to achieve mobility (Clarke et al. 2018). At the end of the 5-year bond, the five outcome donors will pay investors an amount contingent upon the efficiency of the three centers which will then be verified by a third-party independent auditor using a benchmark stipulated in the contract (Carè and Ferraro 2019). This is based on an outcome-based staff efficiency ratio which is the number of beneficiaries gaining mobility per local rehabilitation professional at the end of the intervention (Alenda-Demoutiez 2020). The HIB's private investors consisted of a consortium of nine investors coordinated by Swiss private bank Lombard Odier. The total upfront investment for the bond was USD$19.7 million and the total outcome funding was USD$27.6 million (Clarke et al. 2018).

From a financial perspective, the bond structure is based on a "pay-for-success" type model whereby investors provide the upfront capital needed for the program, investors then receive a payment if the results of the program are above the benchmark specified in the contract, and if on the other hand the results are below the benchmark, then investors will lose part of their capital (Monnet and Panizza 2017). If outcomes are achieved, investors will receive 100% of their principal investment. Moreover, if the program achieves outcomes above a pre-agreed target level, then investors will receive their principal investment plus a financial return of up to 7% interest, and if outcomes are not achieved then investors can lose up to 40% of their original investment and all of their interest payments (Clarke et al. 2018).

As a service provider, the ICRC faces some downside financial risk. A staff efficiency ratio of the three new centers built will be compared to an average calculated from existing centers with similar qualities in their second year of operation, if the staff efficiency ratio of the new centers does not reach 100% of the level of "the baseline average," then the ICRC will have to repay investors a proportion of their upfront investment (KOIS Invest 2017). Furthermore, the ICRC will have to pay 10% of any investment lost if minimum targets are not met, but if targets are met or exceeded, the ICRC will not receive any bonus payments (Clarke et al. 2018).

Analysis

There are several key points that the issuance of the HIB raises, these will be discussed in turn below.

Only Funding Activities with Measurable Outcomes
Performance-based contract outcomes tend to relate to the use or coverage of specific and easily measurable health services for example, the share of children who are immunized, the number of pregnant women who are seeking prenatal care or the amount of deliveries that take place at a facility etc.... (Dupas and Miguel 2017). One of the potential risks of contracting over such narrow and specific metrics is, service providers may expend too much effort on such activities rather than health outcomes that may be difficult or expensive to measure. Moreover, with such innovative forms of financing, there is the potential risk of most funds being allocated to activities with clear measurable outcomes or where the likelihood of success is very high however, not all outcomes, especially in the humanitarian sector, are easily measureable (Monnet and Panizza 2017).

In the case of the HIB, there were two main outcomes, first the opening of three new physical rehabilitation centers and second, the improvement in staff efficiency ratio which is how many people receive mobility devices per physical rehabilitation professional which is then benchmarked to existing centers. The first outcome is easily measurable where the project is considered a "success" if the new centers are built and the likelihood of success of this metric is high. However, for the second outcome, a few potential issues could arise. The second outcome is calculated by the number of beneficiaries having regained mobility due to mobility devices being provided by the ICRC, divided by the number of local rehabilitation professions. This will then be compared to the baseline staff efficiency ratio from data collected from comparable ICRC centers in Africa. Although this metric does incorporate an element of quality, there is a risk of focusing on quantity to better meet targets. Furthermore, the ratio may introduce the risk of service providers "cherry picking" beneficiaries in order to reach targets so the project is deemed a success.

In developing countries prevailing attitudes on disability mean that disability is still treated as either a medical issue, with a focus on initiatives to prevent impairments or as a charity issue, with interventions designed on the assumption that long-term welfare assistance should be the main response to meet the needs of people living with disabilities (Lang and Upah 2008). Although work conducted on these prevailing attitudes has brought benefits to disabled persons', the underlying assumptions of such interventions are grounded in the view that disabled persons are unproductive, dependent entities that need to be "mitigated" or "dealt with" (Coe and Wapling 2010). This is different from the social model of disability that focuses on the social exclusion of disabled persons (Barnes and Oliver 1993) and recognizes disabled persons as integral members of society. Disability can therefore be seen as a social consequence of a person having an impairment in other words, it is society, and not the impairment itself, that disables through barriers that are social, cultural, economic or environmental (Hurst and Albert 2006).

According to Coe and Wapling (2010), the challenge then is for humanitarian organizations to apply the social model to its work in countries where the medical and charity approach is prevalent and to acknowledge the social model approach as appropriate and necessary to tackle the poverty faced by millions of people living with disabilities worldwide. In regard to the HIB, it fits within the conventional medical/charity approach focusing on rehabilitation as it is probably an easier measurable outcome and the likelihood of success is high. The social model in this case would be harder to implement as it could be perceived as riskier or harder to evaluate. Nevertheless, a shift in perspective might allow for moving beyond the charity/medical model to the social model approach to challenge the societal and economic barriers excluding those with disabilities from fully participating in society.

Potential Manipulation in Recorded Outcomes
One of the risks associated with impact bonds is the issue of performance management itself which is one of the pillars of HIBs. The literature concerning performance-based financing and results-based financing mechanisms outlines various biases that are as a result of short-term thinking rather than long-term visions such as distortions between target audiences and others, the cherry-picking of beneficiaries, the focus on quantity rather than quality, the increase in inequality by rewarding service providers and facilities that are better positioned to meet targets (Ireland et al. 2011). These same concerns could be applicable to SIBs which could lead service providers "to focus their activity on meeting whichever indicators are measured, at the potential expense of other, perhaps more important, issues, not included among metrics" (Roy et al. 2017).

Moreover, rewarding performance for specific outcomes may cause those delivering services such as frontline health workers, to re-optimize services in a way that maximizes those outputs while hindering the overall health outcomes of patients (Belinsky et al. 2014). For example, pressures to meet targets could result in unintended consequences whereby service providers are incentivized to achieve good patient outcomes so they "cherry pick" patients who are the healthiest or easiest to treat versus those who are more sick or located in remote areas (Miller and Babiarz 2014). Incentivizing specific services could also lead to other consequences such as motivating the fabrication of performance evaluation sheets (Ellingsen and Johannesson 2008) for example, a maternal health program implemented in India started paying staff more for babies delivered after office hours as a result, the number of night-time deliveries sharply increased which indicated that staff falsified results in order to receive additional payments (Vora et al. 2009).

According to Belinsky et al. (2014), healthcare impact bonds must be designed carefully to ensure unwanted behavior is not motivated for instance, extrinsic motivation, such as that presented by monetary incentives provided to frontline health workers which may crowd out intrinsic motivation such as an altruistic desire to help patients. A review of pay-for-performance incentives of health programs in low- and middle-income countries found "financial incentives may lead to demoralization, reductions in intrinsic motivation, less trust between patients and providers" and may ultimately decrease the quality of the healthcare workforce in the long run by selecting against individuals who are motivated to perform well intrinsically (Miller and Babiarz 2014).

Leveraging Private Capital

Is has been suggested that one of the advantages of using impact bonds to fund social interventions is, risk is shifted from the public to the private sector as investors are only repaid and receive a return on their investment if the program achieves its stipulated objectives (Dowling 2017). However, if the program does not achieve its outcomes then investors not only lose their principal but they also do not receive a return. According to advocates of SIBs, social investment is about "financing the unbankable" (Social Investment Research Council 2015), in other words, bringing organizations and initiatives that would not typically receive private finance or be considered too risky into the realms of finance.

Proponents of impact bonds also suggest that the introduction of private social investment in payment by results contracts has the ability to create experimentation in service and innovation as it redistributes part of the financial risk of non-delivery away from public sector entities and small providers to social investors (Disley et al. 2011; Cooper et al. 2013). In theory, it means that it is possible for smaller and civil society organizations to participate in outcomes-based commissioning in a way that they have not done so or struggled to do so (Edmisn and Aro 2016). Impact bonds are repeatedly referenced as a means to "promote innovation in social services and bring market forces to bear on service providers previously funded by traditional government grants" (Cooper et al. 2013); furthermore, public sector and social finance stakeholders such as social investors and social impact investment entities have stated that the presence of private actors has enabled third sector organizations to foster innovation in service delivery (Edmiston and Aro 2016).

Funding raised from the HIB is allocated toward building physical rehabilitation centers and essentially provides mobility devices for disabled persons. Persons with disabilities are one of the most vulnerable and socially excluded groups in displaced communities in addition, barriers to accessing humanitarian assistance programs increase their protection risks including risk of abuse, exploitation and violence (Pearce 2015). The relationship between economic inequalities and disablement has been well established given the cyclical correlation between poverty and disability (WHO 2011); however, there has been less focus on how that impacts social cohesion over time and affects disabled people inter-generationally (Berghs 2015).

In this case although the HIB is innovative in form, by design it does not appear that innovative as the ICRC have used proceeds raised from the impact bond to expand their existing physical rehabilitation program rather than experiment and innovate with the type of services they deliver. For example, funding could have been allocated toward social cohesion programs for disabled persons which is vastly underfunded given that disability is predominantly treated as a medical issue rather than a social one in humanitarian contexts. Mainstreaming disability during and post conflict could foster and create a more equitable society altering the conditions that ignite violence and conflict in the first place (Berghs 2015).

While it is well known that disabled persons in the Global South are disproportionately impacted by humanitarian crises, they are rarely consulted about their needs (Twigg 2014). According to Berghs (2015), ensuring the links between health, rehabilitation, justice and preventative services is essential in dealing with a rising group of people with physical impairments and mental health trauma which is difficult as disability is often seen as a specialized medical humanitarian issue. Instead, disability needs to be seen as a cross-cutting issue that can contribute to greater social justice (Berghs 2015). Given the focus on measurable outcomes, although in theory, impact bonds are supposed to fund riskier social programs and initiatives in reality, designing a HIB to tackle the social issues associated with disabilities and mainstreaming disability to provide positive socioeconomic impacts in humanitarian contexts is more difficult to quantify and therefore measure.

Creating Sustainable Funding
Humanitarian grant financing tends to be provided in either annual or even shorter cycles, despite the nature of the crisis, which is not conducive to appropriate financial planning. Furthermore, the scope and potential of programs and initiatives are at risk due to the lack of consistent, predictable multi-year funding. The ICRC has increasingly been taking an approach whereby humanitarian staff work toward both a short- and long-term timeframe in their programs for instance, merging elements such as immediate food relief with seed supply, livestock vaccinations with cash distribution and emergency medical care with the training of hospital staff and guaranteeing their incomes (ICRC 2016a, b).

The HIB is an innovative instrument in the sense that it creates incentives to fund programs over a longer period of time, in this case 5 years, allowing humanitarian organizations adequate time for planning. As one of the outcomes for the bond is the construction of rehabilitation centers, this type of multi-year funding is ideal as infrastructure financing requires funding over a much longer period of time due to its long-gestation nature. However, 5 years is still not an adequate timeframe to provide rehabilitation services for disabled persons and if the support to prosthetic services is reduced or removed without any replacement due to a lack of funding, then it would have an overwhelmingly negative impact on a group that is already vulnerable.

According to the Centre for Global Development's Development Impact Bond Working Group, impact bonds can shift the paradigm of how social programs are funded by creating sustained funding (Clarke et al. 2018). Evaluating the HIB based on this criterion to assess whether the HIB creates a sustainable financing model or source of funding is difficult to ascertain. Although on the one hand the bond appears to be over a longer time period than traditional grant financing, it still does not differ drastically from multi-year donor funding in terms of timeframe. Furthermore, the bond appears to be another form of earmarked funding with the exception of the addition of private investors—this type of humanitarian aid refers to donor funding contributions to organizations specifically earmarked for certain purposes including regions, countries, themes or sectors (OECD 2011). Over the last two decades, donor governments have increased the level of foreign aid provided to humanitarian organizations as earmarked funding (Eichenauer and Reinsberg 2017). However, earmarked funding decreases the flexibility and agency of humanitarian organizations in determining their expenditures.

Bourgeoning Humanitarian–Development Nexus
The case also demonstrates how traditional approaches to humanitarian aid are changing as more humanitarian organizations transcend the line between humanitarian aid and development cooperation. The aid system has traditionally been partitioned between humanitarian and development assistance. Humanitarian aid is geared toward addressing emergency situations to meet the immediate needs of those impacted by crisis while development aid seeks to address more structural causes of poverty by working to change the social, economic and political systems that create the conditions in which poverty and inequality exists (Bennett 2015; Brown and Donini 2014; Kocks et al. 2018).

Humanitarian actors are guided by four principles which are humanity, impartiality, independence and neutrality, which provides them with the moral obligation to distribute aid to anyone in need regardless of their affiliations, without the influence from the interest of both political and non-political actors, and with no purpose or interest to influence the outcome of a conflict (Chandler 2001). Whereas the development sector is not meant to be neutral, impartial or independent as the humanitarian sector in its approach, instead its work is rooted in a human rights framework that aims to uphold the rights of the beneficiaries of aid in accordance with relevant bodies of law (Brown and Donini 2014; CIC 2015).

Funding for humanitarian assistance is therefore typically short-term in nature in order to be flexible and be able to respond quicker to deliver aid for those in crisis contexts. While on the other hand, development assistance is longer-term and therefore provides some stability, predictability and dependability for all stakeholders involved (Gavas et al. 2015; Kocks et al. 2018; Watson 2016; UNSG 2016). Furthermore, development aid focuses on a different population than humanitarian aid as development initiatives target citizens of the countries in which they work and tends to work through national systems (Hinds 2015; Macrae 2012). As such, development organizations aim to be sustainable in their efforts by working with local and national government structures (Kaga and Nakache 2019).

The case of transition from immediate humanitarian action to development aid has illustrated significant changes occurring in the humanitarian sector giving momentum to what has been described as an ongoing "humanitarian mission creep" (Barnett 2011) through which humanitarian organizations expand their apolitical scope by moving beyond offering "a bed for the night" (Rieff 2003), and more toward widening humanitarian assistance into prevention and recovery efforts.

After they are built using the proceeds raised from the HIB, the rehabilitation centers will be managed by the national health authorities of Mali, Nigeria and the Democratic Republic of the Congo and for the first three years, the ICRC will train local staff in the use of equipment, producing prostheses and providing rehabilitation and thereafter will continue to support the centers as needed indefinitely (Devex 2017). This is indicative of the wider trend of the burgeoning humanitarian-development nexus. As protracted crisis become the "new normal," humanitarian organizations have needed to expand their operations and scope of work into recovery and development provision.

Although this demonstrates that evolving humanitarian funding models are needed to address more complex humanitarian issues, a few issues arise. As humanitarian organizations expand their operations into the development sector, this could put at risk the fundamental humanitarian principles of humanity, impartiality, neutrality and in particular, independence. Moreover, this also raises concerns about the infringement of state sovereignty as humanitarians work beyond their typical role in a crisis. The relationship between humanitarian actors and states has long been dynamic and complex with humanitarian organizations being criticized for undermining and hindering states role as providers for their citizens (Kahn and Cunningham 2013). However, as humanitarian organizations expand into the development sector using innovative funding models such as the HIB, this could be an indicator of a wider trend which is the rise of the private sector's role, and potential influence, on states even though they may not share the same principles, ideological perspectives or goals.

Financialization of Aid

The use of impact bonds such as HIBs in the humanitarian sector also brings to light some political and ideological questions. Impact bonds such as SIBs have been viewed as a further sign of the growing influence of market and financial logics over that of social services, it is seen as the first step toward a "slippery slope" to the financialization of social services (Lake 2016; Sinclair et al. 2019; Tse and Warner 2020; Warner 2013). Furthermore, according to some critics, SIB models represent a paradigm shift toward the privatization and marketization of the social sector (Williams 2019).

In a broad sense, financialization is the "increasing role of financial motives, financial markets, financial actors and financial institutions in the operation of the domestic and international economies" (Epstein 2005). Financialization can create new patterns of capital accumulation (Fine 2012; Krippner 2005) and shape social establishments and subjectivities (Davis and Kim 2015; van der Zwan 2014), leading to new forms of social regulation (Storm 2018) and resistance (Davis and Kim 2015). The financialization–development nexus has been scrutinized in recent years (Mawdsley 2018) as researchers have indicated a dissociation of financial returns from the productive economy (Fine 2012), the cannibalization of the productive economy in order to serve the interests of finance (Storm 2018) and the prioritization of shareholder concerns over other issues such as social impact (Hunter and Murray 2019).

Beyond the results of the HIB, the program raises questions as to what the implications are of using such an instrument to carry out a humanitarian response. For instance, the benevolent actions of investors could be dismissed as they could earn financial returns as well as interest payments, rather than them challenging the distinct social and economic implications of those with disabilities in humanitarian settings. In other words, to what extent do investors consider delivering greater social impact of the program or the social good created rather than just the financial return, given impact bonds stress the achievement of measurable, quantifiable outcomes?

Although the HIB as a funding instrument is perceived as innovative and helping to plug the immense humanitarian funding gap, whether it can lessen the burden of social provision on humanitarian organizations is yet to be seen.

Conclusion

This chapter reviewed HIBs as a nascent part of the wider set of ideas and practices that are gaining popularity in contemporary humanitarianism.

In general, HIBs are not only considered as innovative financial instruments but also represent a boundary and ideological shift in humanitarian provision that has been challenging traditional practices of humanitarian assistance by disciplining the organizations that administer and provide it (Krause 2014). Furthermore, the trend toward a stronger involvement of private sector actors in the humanitarian sector can be linked to the new discourse on "social entrepreneurship" that emphasizes a triple bottom-line with a focus on social problem-solving as well as financial and environmental sustainability (Nicholls 2006; Trexler 2008; Guo and Bielefeld 2014).

As a financial instrument, HIBs creates a new opportunity for private sector actors such as financial institutions and investors, to make social and financial commitments and reimagine the humanitarian sector based on ideals of improving outcomes. This ideological shift is not limited to the provision of humanitarian assistance but is a paradigm move that can be delineated in various debates around the notions of effective altruism in philanthropy (Gabriel 2017), results-based financing to achieve the United Nations sustainable development goals (Gustafsson-Wright et al. 2017) and plugging the humanitarian financing gap (High-Level Panel on Humanitarian Financing 2016). However, some concerns have arisen about the current shift toward private sector finance and investment in the humanitarian sector with the potential risk of diverting attention away from the core humanitarian principles of humanity, independence, impartiality and neutrality and toward more measurable "successful" projects based on outcomes-based performance management mechanisms.

In order to assess the implications of such instruments for the practices and principles of humanitarianism, this chapter has engaged a case study of the ICRC HIB, which was the first of its kind. It is found that in the context of this program, this new mode of financing sought to address current challenges of humanitarian assistance provision through a results culture and focus on measurable socially-oriented outcome targets and performance management. The "new normal" of humanitarian financing is thus a direct concern of this chapter, reflecting the way in which humanitarian actors weigh humanitarian principles with financial objectives but also how HIBs represent a results-oriented culture for service providers incentivized to strive for better outcomes which can impact the operation of interventions. Whether HIBs can shift the culture of how humanitarian programs are funded is yet to be seen.

Although the case study method is useful in being able to draw inferences, this method is limited in part because this project alone is not a complete reflection of this new type of direction in funding humanitarian programs. Furthermore, the ICRC HIB is yet to be completed; therefore, further assessment and scrutiny of the program is needed ex post to examine the results. Research that further examines the various actors, processes, projects and effects of the use of impact bonds will be key to unravelling the pervasiveness of the narrative of a strictly results-oriented culture and performance management approach to be applied to humanitarian contexts and issues.

References

Addis, R., McLeod, J., and Raine, A. (2013). *Impact—Australia: Investment for social and economic benefit*. Canberra: Department of Education, Employment and Workplace Relations.

Albertson, K., and Fox, C. (2018). *Payment by results and Social Impact Bonds: Outcome-based payment systems in the UK and US*. Bristol, England: Policy Press.

Alenda-Demoutiez, J. (2020). A fictitious commodification of local development through development impact bonds? *Journal of Urban Affairs*, 42(6), 892–906.

Andreu, M. (2018). A responsibility to profit? Social impact bonds as a form of "humanitarian finance". *New Political Science*, 40(4), 708–726.

Arena, M., Bengo, I., Calderini, M., and Chiodo, V. (2016). Social Impact Bonds: Blockbuster or flash in a pan? *International Journal of Public Administration*, 39(12), 927–939.

Barnes, C., and Oliver, M. (1993). *Disability: A sociological phenomenon ignored by sociologists*. University of Leeds.

Barnett, M. (2011). *Empire of humanity: A history of humanitarianism*. Ithaca: Cornell University Press.

Bassen, A., and Kovacs, A. (2008). Environmental, social and governance key performance indicators from a capital market perspective. *Zeitschrift für Wirtschaftsund Unternehmensethik (zfwu)*, 9(2), 182–192.

Belinsky, M., Eddy, M., Lohmann, J., and George, M. (2014). The application of social impact bonds to universal health-care initiatives in South-East Asia. *WHO South-East Asia Journal of Public Health*, 3(3–4), 219–225.

Bennett, C. (2015). *The development agency of the future: Fit for protracted crises?* Working Paper. Overseas Development Institute. Retrieved from: https://www.odi.org/sites/odi.org.uk/files/odi-assets/publications-opinion-files/9612.pdf.

Berghs, M. (2015). Radicalising 'disability' in conflict and post-conflict situations. *Disability and Society*, 30(5), 743–758.

Berndt, C., and Wirth, M. (2018). Market, metrics, morals: The Social Impact Bond as an emerging social policy instrument. *Geoforum*, 90, 27–35.

Bertone, M. P., Jacobs, E., Toonen, J., Akwataghibe, N., and Witter, S. (2018). Performance-based financing in three humanitarian settings: principles and pragmatism. *Conflict and Health*, 12(1), 1–14.

Bollag, B. (2017). *ICRC Launches World's First Humanitarian Impact Bond*. Devex.

Bovaird, T. (2014). Attributing outcomes to social policy interventions – 'gold standard' or 'fool's gold' in public policy and management? *Social Policy & Administration*, 48(1), 1–23.

Brien, S. (2011). Outcome-Based Government – How to improve spending decisions across government. London: Centre for Social Justice.

Brown, D., and Donini, A. (2014). *Rhetoric or reality. Putting affected people at the centre of humanitarian action*. London: Active Learning Network for Accountability and Performance (ALNAP), Overseas Development Institute.

Bugg-Levine, A., and Emerson, J. (2011). Impact investing: Transforming how we make money while making a difference. *Innovations: Technology, Governance, Globalization*, 6(3), 9–18.

Cabinet Office. (2015). Centre for Social Impact Bonds: Map of Social Impact Bonds.

Canadian Task Force on Social Finance. (2010). *Mobilizing private capital for public good*. Toronto, ON: Social Innovation Generation.

Carè, R., and Wendt, K. (2018). Investing with impact: An integrated analysis between academics and practitioners. In *Social impact investing beyond the SIB*. Palgrave Macmillan.

Carè, R., and Ferraro, G. (2019). Funding innovative healthcare programs through Social Impact Bonds: Issues and challenges. *China-USA Business Review*, 18, 1–15.

Carroll, A. (1991). The pyramid of corporate social responsibility: Toward the moral management of organizational stakeholders. *Business Horizons*, 34, 39–48.

Center on International Cooperation (CIC). (2015). *Addressing protracted displacement: A framework for development-humanitarian cooperation* (pp. 1–28). New York: New York University.

Chandler, D. (2001). The road to military humanitarianism: How the human rights NGOs shaped a new humanitarian agenda. *Human Rights Quarterly*, 23(3), 678–700.

Chiapello, E. (2015). Financialisation of valuation. *Human Studies* 38(1), 13–35.

Clarke, L., Chalkidou, K., and Nemzoff, C. (2018). *Development impact bonds targeting health outcomes*. Center for Global Development, Policy

Paper, 133. Retrieved from: https://www.cgdev.org/sites/default/files/dev elopment-impact-bonds-targeting-health-outcomes.pdf.

Coe, S., and Wapling, L. (2010). Practical lessons from four projects on disability-inclusive development programming. *Development in Practice*, 20(7), 879–886.

Cooper, C., Graham, C., and O'Dwyer, B. (2013). *Social impact bonds: Can private finance rescue public programs?* Paper presented at Accounting, Organizations and Society Conference on "Performing Business and Social Innovation through Accounting Inscriptions". Galway, Ireland.

Council on Social Action. (2008). *Council on Social Action: Commentary on year one*. United Kingdom: Community Links.

Davis, G.F. and Kim, S. (2015). Financialization of the economy. *Annual Review of Sociology*, 41(1): 203–221.

Dear, A., Helbitz, A., Khare, R., Lotan, R., Newman, J., Sims, G. C., and Zaroulis, A. (2016). *Social Impact Bonds: The early years*. United Kingdom: Social Finance.

Disley, E., Rubin, J., Scraggs, E., Burrowes, N., and Culley, D. (2011). Lessons learned from the planning and early implementation of the Social Impact Bond at HMP Peterborough. *Research Series*, 5(11). Cambridge: RAND Europe.

Dowling, E. (2017). In the wake of austerity: Social Impact Bonds and the Financialisation of the Welfare State in Britain. *New Political Economy*, 22(3), 294–310.

Drexler, M., Noble, A., and Bryce, J. (2013). *From the margins to the main-stream: Assessment of the Impact Investment Sector and Opportunities to Engage Mainstream Investors*. Geneva, Switzerland: World Economic Forum.

Dupas, P., and Miguel, E. (2017). Impacts and determinants of health levels in low-income countries. In *Handbook of economic field experiments*. North-Holland.

Edmiston, D., and Aro, J. (2016). *Public policy, social innovation and marginali-sation in Europe: A comparative analysis of three cases*. CRESSI Working Paper Series No. 33/2016. Oxford: University of Oxford.

Eichenauer, V. Z., and Reinsberg, B. (2017). What determines earmarked funding to international development organizations? Evidence from the new multi-bi aid data. *The Review of International Organizations*, 12(2), 171–197.

Eide, A. H. (2013). Users of the physical rehabilitation services supported by ICRC Special Fund for the disabled in Vietnam. In *Description and assessments of impact and future service needs*. Oslo: SINTEF Technology and Society.

Ellingsen, T., and Johannesson, M. (2008). Pride and prejudice: The human side of incentive theory. *American Economic Review*, 98(3), 990–1008.

Epstein, G. A. (Ed.). (2005). *Financialization and the world economy*. Cheltenham: Edward Elgar Publishing Ltd.

Fama, E. F., and Jensen, M. C. (1983). Separation of ownership and control. *Journal of Law and Economics*, 26(2), 301–325.

Finance for Good. (2015). *Finance for good social impact bond tracker.* Retrieved from: http://financeforgood.ca/social-impact-bond-resources/sib-tracker/.

Fine, B. (2012). Neoliberalism in retrospect? It's financialisation, stupid. In *Developmental politics in transition: The neoliberal era and beyond.* London: Palgrave Macmillan.

Fitzgerald, J. L. (2013). Social impact bonds and their application to preventive health. *Australian Health Review*, 37(2), 199–204.

Floyd, D. 2017. *Social Impact Bonds, an overview of the global market for commissioners and policymakers.* Social Spider CIC & Centre for Public Impact: A BCG Foundation. Retrieved from: http://socialspider.com/wp-content/uploads/2017/04/SS_SocialImpactReport_4.0.pdf.

Fox, C. and Albertson, K. (2011). Payment by results and social impact bonds in the criminal justice sector: New challenges for the concept of evidence based policy? *Criminology and Criminal Justice*, 11(5), 395–413.

Freeman, R. E. (1984). *Strategic management: A stakeholder approach.* Boston: Pitman.

Gabriel, I. (2017). Effective altruism and its critics. *Journal of Applied Philosophy*, 34(4), 457–473.

Galbreath, J. (2013). ESG in focus: The Australian evidence. *Journal of Business Ethics*, 118(3), 529–541.

Gavas, M., Gulrajani, N., and Hart, T. (2015). Designing the development agency of the future. In *Framing paper for the conference 'Designing the development agency of the future'.* London: Overseas Development Institute (ODI).

Geobey, S., Westley, F. R., and O. Weber. (2012). Enabling social innovation through developmental social finance. *Journal of Social Entrepreneurship*, 3(2): 151–165.

Giacomantonio, C. (2017). Grant-maximizing but not money-making: A simple decision-tree analysis for social impact bonds. *Journal of Social Entrepreneurship*, 8(1), 47–66.

Global Impact Investing Network (GIIN). (2017). *Annual Impact Investor Survey.* Retrieved from: https://thegiin.org/assets/GIIN_AnnualImpactInvestorSurvey_2017_Web_Final.pdf.

Government Outcomes Lab. (2018a). *An introduction to Development Impact Bonds (DIBs).* Retrieved from: https://golab.bsg.ox.ac.uk/knowledge/basics/introduction-developmentimpact-bonds/. University of Oxford.

Government Outcomes Lab. (2018b). *Humanitarian Impact Bond.* University of Oxford. Retrieved from: https://golab.bsg.ox.ac.uk/knowledge/casestudies/humanitarian-impact-bond/.

Griffiths, A., and Meinicke, C. (2014). *Introduction to Social Impact Bonds and early intervention—Initial report*. Early Intervention Foundation.

Guo, C., and Bielefeld, W. (2014). *Social entrepreneurship: An evidence-based approach to creating social value*. San Francisco, CA: Jossey-Bass.

Gustafsson-Wright, E., Boggild-Jones, I., Segell, D., and Durland, J. (2017). *Impact bonds in developing countries: Early learning from the field*. Washington, DC: Center for Universal Education at Brookings; Convergence. Retrieved from: https://www.brookings.edu/wp-content/uploads/2017/09/impact-bonds-in-developing-countries_web.pdf.

Harji, K., and Jackson, E. (2012). *Accelerating impact: Achievements, challenges and what's next in building the impact investing industry*. New York: The Rockefeller Foundation.

Hedderman, C. (2013). Payment by results: Hopes, fears and evidence. *British Journal of Community Justice*, 11(2–3), 43–58.

High-Level Panel on Humanitarian Financing. (2016). *Too important to fail—Addressing the humanitarian financing gap*. New York, NY: United Nations.

Hinds, R. (2015). *Relationship between humanitarian and development aid. Governance and Social Development Resource Centre (GSDRC)*. Birmingham, UK: University of Birmingham.

Höchstädter, A. K., and Scheck, B. (2015). What's in a name: An analysis of impact investing understandings by academics and practitioners. *Journal of Business Ethics*, 132(2), 449–475.

Howard, N., Woodward, A., Souare, Y., Kollie, S., Blankhart, D., von Roenne, A., and Borchert, M. (2011). Reproductive health for refugees by refugees in Guinea III: Maternal health. *Conflict and Health*, 5, 1–8.

Humphries, K. W. (2013). Not your older brother's bonds: The use and regulation of social-impact bonds in the United States. *Law & Contemporary Problems*, 76, 433–452.

Hunter, B. M., and Murray, S. F. (2019). Deconstructing the financialization of healthcare. *Development and Change*, 50(5), 1263–1287.

Hurst, R., and Albert, B. (2006). The social model of disability: Human rights and development cooperation. In *Lessons from research on disability and development cooperation*. Leeds: The Disability Press, 24–39.

International Committee of the Red Cross (ICRC). (2016). *Protracted conflict and humanitarian action: Some recent ICRC experiences*. Geneva: ICRC.

International Committee of the Red Cross. (2016). *Physical Rehabilitation Programme: Annual Report*. Retrieved from: https://shop.icrc.org/physical-rehabilitation-programme-annual-report-2016-en-pdf.

Ireland, M., Paul, E., and Dujardin, B. (2011). Can performance-based financing be used to reform health systems in developing countries? *Bulletin of the World Health Organization*, 89, 695–698.

Jacobs, M., and Mazzucato, M. (Eds.). (2016). *Rethinking capitalism: Economics and policy for sustainable and inclusive growth.* John Wiley & Sons.

Jasch, C. (2006). Environmental management accounting (EMA) as the next step in the evolution of management accounting. *Journal of Cleaner Production,* 14(14), 1190–1193.

Kaga, M., & Nakache, D. (2019). *Protection and the humanitarian-development nexus: A literature review.*

Kahn, C., and Cunningham, A. (2013). Introduction to the issue of state sovereignty and humanitarian action. *Disasters,* 37, 139–150.

Kasper, G., and Marcoux, J. (2014). The re-emerging art of funding innovation. *Stanford Social Innovation Review,* 12(2), 28–35.

Khatib, I. M., Samrah, S. M., & Zghol, F. M. (2010). Nutritional interventions in refugee camps on Jordan's eastern border: Assessment of status of vulnerable groups. *East Mediterranean Health Journal,* 16(2), 187–193.

Kocks, A., Wedel, R., Roggemann, H., and Roxin, H. (2018). Building bridges between international humanitarian and development responses to forced migration: A review of conceptual and empirical literature with a case study on the response to the Syria crisis. MISC.

KOIS Invest. (2017). *Special Edition Newsletter—Launching the World's First Humanitarian Impact Bond with the ICRC.* Retrieved from: https://docs. wixstatic.com/ugd/679693_99c86289da4f4a1b9d0849ad1b9d7a85.pdf.

Krause, M. (2014) *The good project: Humanitarian relief NGOs and the fragmentation of reason.* Chicago, IL: University of Chicago Press.

Krippner, G. R. (2005). The financialization of the American Economy. *Socio-Economic Review,* 3(2), 173–208.

La Torre, M., and Calderini, M. (Eds.). (2018). *Social impact investing beyond the SIB: Evidence from the market.* London: Palgrave Macmillan.

Lake, Robert. (2016). The subordination of urban policy in the time of financialization. In J. DeFillippis (Eds.), *Urban social policy in the time of Obama.* Minneapolis: University of Minnesota Press.

Lang, R., and Upah, L. (2008). *Scoping study: Disability issues in Nigeria.* United Kingdom: Department for International Development (DfID).

Lehner, O. M. (Ed.). (2016). *Routledge handbook of social and sustainable finance.* London: Routledge.

Louche, C., Arenas, D., and Van Cranenburgh, K. C. (2012). From preaching to investing: Attitudes of religious organisations towards responsible investment. *Journal of Business Ethics,* 110(3), 301–320.

Lowe, T. (2013). New development: The paradox of outcomes—The more we measure, the less we understand. *Public Money & Management,* 33(3), 213–216.

Lowe, T., and Wilson, R. (2017). Playing the game of outcomes-based performance management. Is gamesmanship inevitable? Evidence from theory and practice. *Social Policy & Administration*, 51, 981–1001.

Macrae, J. (2012). The continuum is dead, long live resilience. *VOICE Out Loud*, 15, 8–10.

Mawdsley, E. (2018). Development geography II: Financialization. *Progress in Human Geography*, 42(2): 264–274.

Meekosha, H. (2011). Decolonizing disability: Thinking and acting globally. *Disability & Society*, 26(6), 667–682.

Miller, G., and Babiarz, K. S. (2014). Pay-for-performance incentives in low- and middle-income country health programs. In *Encyclopedia of health economics* (Vol. 2). San Diego, CA: Elsevier.

Mirza, M. (2015). Disability-inclusive healthcare in humanitarian camps: Pushing the boundaries of disability studies and global health. *Disability and the Global South*, 2(1), 479–500.

Mitchell, K. (2017). 'Factivism': A new configuration of humanitarian reason. *Geopolitics*, 22(1), 110–128.

Mollica, R. F., Donelan, K., Tor, S., Lavelle, J., Elias, C., Frankel, M., and Blendon, R. J. (1993). The effect of trauma and confinement on functional health and mental health status of Cambodians living in Thailand-Cambodia border camps. *JAMA: Journal of the American Medical*, 270(5), 581–586.

Monnet, N., and Panizza, U. (2017). *A note on the economics of philanthropy*. Graduate Institute of International and Development Studies Working Paper, 19. Retrieved from: https://repository.graduateinstitute.ch/record/295337?ln=en.

Moore, M. L., Westley, F. R., and Nicholls, A. (2012). The social finance and social innovation nexus. *Journal of Social Entrepreneurship*, 3(2), 115–132.

National Audit Office. (2015). *Outcome-based payment schemes: Government's use of payment by results*.

Nicholls. A. (Ed.). (2006). *Social entrepreneurship: New models of sustainable social change*. Oxford: Oxford University Press.

Nicholls, A. (2010a). The institutionalization of social investment: The interplay of investment logics and investor rationalities. *Journal of Social Entrepreneurship*, 1(1), 70–100.

Nicholls, A. (2010b). The functions of performance measurement in social entrepreneurship: Control, planning and accountability. In *Values and opportunities in social entrepreneurship*. New York, NY: Palgrave MacMillan.

Nicholls, A., and Tomkinson, E. (2013). *The Peterborough Pilot: Social Impact Bond*. Said Business School, University of Oxford.

O'Donohoe, N., Leijonhufvud, C., Saltuk, Y., Bugg-Levine, A., and Brandenburg, M. (2010). *Impact investing: An emerging asset class*. JP Morgan, The Rockefeller Foundation and Global Impact Investing Network.

OECD. (2011). *DAC report on multilateral aid*. DCD/DAC 21/FINAL. Paris: Organization for Economic Cooperation and Development (OECD).

Oroxom, R., Glassman, A., McDonald, L. (2018). *Structuring and funding development impact bonds for health: Nine lessons from Cameroon and beyond*. Washington, DC: Center for Global Development, Policy Paper No. 117.

Pearce, E. (2015). 'Ask us what we need': Operationalizing guidance on disability inclusion in refugee and displaced persons programs. *Disability and the Global South*, 2(1), 460–478.

Perrin, B. (1998). Effective use and misuse of performance measurement. *American Journal of Evaluation*, 19(3), 367–379.

Porter, M. E., and Kramer, M. R. (2011). Creating shared value: How to reinvent capitalism—And unleash a wave of innovation and growth. *Harvard Business Review*, 89(1–2), 62–77.

Ragin, L., and Palandjian, T. (2013). Social impact bonds: Using impact investment to expand effective social programs. *Community Development Investment Review*, 63–67.

Rangan, V. K., and Chase, L. A. (2015). The payoff of pay-for-success. *Stanford Social Innovation Review*, 4, 28–39.

Renmans, D., Holvoet, N., Criel, B., and Meessen, B. (2017). Performance-based financing: the same is different. *Health Policy and Planning*, 32(6), 860–868.

Rieff, D. (2003). *A bed for the night: Humanitarianism in crisis*. London: Simon and Schuster.

Rizzello, A., Migliazza, M., Care, R., and Trotta, A. (2016). Social impact investing: A model and research agenda. In *Routledge handbook of social and sustainable finance*. London: Routledge.

Roca, M. G., Charle, P., Jiménez, S., & Núñez, M. (2011). A new malaria protocol in a Congolese refugee camp in West Tanzania. *Global Public Health*, 6(4), 398–406.

Roy, M., McHugh, N., and Sinclair, S. (2017). Social impact bonds: Evidence-based policy or ideology? In *Handbook of social policy evaluation*. Cheltenham, England: Edward Elgar.

Saltuk, Y. (2011). Counter(imp)acting austerity: The global trend of government support for impact investment. New York, NY: J.P. Morgan.

Schalock, R. (2001). *Outcome based evaluation*. New York: Kluwer Academic.

Schinckus, C. (2015). The valuation of social impact bonds: An introductory perspective with the Peterborough SIB. *Research in International Business and Finance*, 35, 104–110.

Schinckus, C. (2017). Financial innovation as a potential force for a positive social change: The challenging future of social impact bonds. *Research in International Business and Finance*, 39, 727–736.

Shiller, R. J. (2013). *Finance and the good society*. Princeton University Press.

Shivji, A. (2010). Disability in displacement. *Forced Migration Review*, 35, 4–7.
Sila, I., and Cek, K. (2017). The impact of environmental, social and governance dimensions of corporate social responsibility on economic performance: Australian evidence. *Procedia Computer Science*, 120, 797–804.
Sinclair, S., McHugh, N., and Roy, M. J. (2019). Social innovation, financialisation and commodification: A critique of social impact bonds. *Journal of Economic Policy Reform*, 1–17.
So, I., and Jagelewski, A. (2013). *Social Impact Bond Technical Guide for Service Providers. MaRS Centre for Impact Investing*. Retrieved from: http://www.marsdd.com/wpcontent/uploads/2013/11/MAR-SIB6939__Social-Impact-Bond-Technical-Guide-for-ServiceProviders_FINAL-ELECTRONIC1.pdf.
Social Impact Investing Taskforce. (2014). *Impact investment: The invisible heart of markets*. Harnessing the power of entrepreneurship, innovation and capital for public good. Retrieved from: https://socialimpactinvestment.org/reports/Impact%20Investment%20Re-port%20FINAL%5B3%5D.pdf
Social Investment Research Council. (2015). *The social investment market through a data lens*. London: Big Society Capital.
Spiegel, P. B., Checchi, F., Colombo, S., and Paik, E. (2010). Health-care needs of people affected by conflict: Future trends and changing frameworks. *The Lancet*, 375, 341–345.
Spiess-Knafl, W., and Scheck, B. (2017). *Impact investing: Instruments, mechanisms and actors*. New York: Palgrave Macmillan.
Stein, M. A., Stein, P. J., Weiss, D., and Lang, R. (2009). Health care and the UN Disability Rights Convention. *The Lancet*, 374, 1796–1798.
Stoesz, D. (2014). Evidence-based policy: Reorganizing social services through accountable care organizations and social impact bonds. *Research on Social Work Practice*, 24(2), 181–185.
Storm, S. (2018). Financialization and economic development: A debate on the social efficiency of modern finance. *Development and Change*, 49(2), 302–329.
Taliento, M., Favino, C., and Netti, A. (2019). Impact of environmental, social, and governance information on economic performance: Evidence of a corporate 'sustainability advantage' from Europe. *Sustainability*, 11(6), 1738.
Tamimi, N., and Sebastianelli, R. (2017). Transparency among S&P 500 companies: An analysis of ESG disclosure scores. *Management Decision*, 55, 1660–1680.
Tarmuji, I., Maelah, R., and Tarmuji, N. H. (2016). The impact of environmental, social and governance practices (ESG) on economic performance: Evidence from ESG score. *International Journal of Trade, Economics and Finance*, 7(3), 67.

Tomlinson M., Swartz, L., Officer, A., Chan, K. Y., Rudan, I., and Saxena S. (2009). Research priorities for health of people with disabilities: An expert opinion exercise. *The Lancet*, 384, 1857–1862.

Trelstad, B. (2009). The nature and type of "social investors". New York, NY: Acumen Fund.

Trexler, J. (2008). Social entrepreneurship as an algorithm: Is social enterprise sustainable? *E:CO* 10(3): 65–85.

Trotta, A., Caré, R., Severino, R., Migliazza, M. C., and Rizzello, A. (2015). Mobilizing private finance for public Good: Challenges and opportunities of Social impact bonds. *European Scientific Journal*, 11, 259–279.

Turban, D. B., and Greening, D. W. (1997). Corporate social performance and organizational attractiveness to prospective employees. *Academy of Management Journal*, 40(3), 658–672.

Tse, A. E., and Warner, M. E. (2020). The razor's edge: Social impact bonds and the financialization of early childhood services. *Journal of Urban Affairs*, 42(6), 816–832.

Twigg, J. (2014). Attitude before method: Disability in vulnerability and capacity assessment. *Disasters*, 38(3), 465–482.

UN Secretary-General (UNSG). (2016). In safety and dignity: Addressing large movements of refugees and migrants. Report of the Secretary-General. UN Doc. A/70/59.

van der Zwan, N. (2014). Making sense of financialization. *Socio-Economic Review*, 12(1), 99–129.

Vora, K. S., Mavalankar, D. V., Ramani, K. V., Upadhyaya, M., Sharma, B., Iyengar, S., Gupta, V. and Iyengar, K. (2009). Maternal health situation in India: A case study. *Journal of Health, Population, and Nutrition*, 27(2), 184.

Warner, M. E. (2013). Private finance for public goods: social impact bonds. *Journal of Economic Policy Reform*, 16(4), 303–319.

Watson, C. (2016). *Financing our shared future: Navigating the humanitarian, development and climate finance agendas*. ODI Briefing. London: Overseas Development Institute (ODI).

Weber, O. (Eds). (2016). Introducing impact investing. In *Routledge Handbook of Social and Sustainable Finance*. Oxford, United Kingdom: Routledge, Taylor Francis Group.

Weber, O., and Feltmate, B. (2016). *Sustainable banking: Managing the social and environmental impact of financial institutions*. Toronto: University of Toronto Press.

Williams, J. W. (2019). *From visions of promise to signs of struggle: Exploring social impact bonds and the funding of social services in Canada, the US, and the UK*. Toronto: York University.

Wilson, K. E. (2014). New investment approaches for addressing social and economic challenges. OECD Science, Technology and Industry Policy Papers,

Oxford. Retrieved from: https://www.bruegel.org/wp-content/uploads/imp orted/publications/OECD_Social_Impact_Investment_overview_paper.pdf.

Witter, S., Fretheim, A., Kessy, F. L., and Lindahl, A. K. (2012). Paying for performance to improve the delivery of health interventions in low-and middle-income countries. *Cochrane Database of Systematic Reviews*, (2).

Wong, G. (2012). *Insights and innovations: A global study of impact investing + institutional investors*. San Mateo, CA: Correlation Consulting.

Wong, M. C. S., and Yap, R. C. Y. (2019). Social impact investing for marginalized communities in Hong Kong: Cases and Issues. *Sustainability*, 11(10), 2831.

Wood, D. (1991). Corporate social performance revisited. *Academy of Management Review*, 16(4), 691–717.

World Health Organization (WHO), and International Disability Development Consortium. (2010). *Community-based rehabilitation: CBR guidelines*. Geneva: World Health Organization (WHO).

World Health Organisation (WHO). (2011). *World report on disability*. Geneva: Switzerland.

Zingales, L. (2015). Presidential address: Does finance benefit society? *Journal of Finance*, 70(4), 1327–1363.

CHAPTER 5

Islamic Social Finance

INTRODUCTION

The Islamic economic and financial system is comprised of a set of laws, rules and principles commonly referred to as "Shari'ah," which is an all-encompassing governance system of the economic, social, political and cultural elements of Muslim societies. Shari'ah is derived and originated from the Holy Quran, the practices of Prophet Mohammed peace be upon him, Islamic jurisprudence through the consensus of religious scholars (Ijma) and personal opinions based on analogy and religious doctrines (Qiyas) (Iqbal and Tsubota 2006). The Muslim individual is therefore guided by a set of behavioral norms derived from the Shari'ah. According to the Kuran (1986), the "primary role of the behavioral norms of Islam is to make the individual member of an Islamic society, homo Islamicus, just, socially responsible and altruistic" whereby the notions of scarcity and self-interest are non-existent by assumption.

The modern form of the Islamic economic system came into existence in the 1970s with the rise of pan-Islamism and the accrual of petro-dollars in some Muslim majority states in an era of high oil prices (Iqbal and Tsubota 2006). The system developed with the objective of combatting the failure of a capitalistic economic development strategy that

© The Author(s), under exclusive license to Springer Nature   101
Switzerland AG 2021
M. Ahmed, *Innovative Humanitarian Financing*,
Palgrave Studies in Impact Finance,
https://doi.org/10.1007/978-3-030-83209-4_5

disregarded the importance of societal well-being in the economic development of Muslim societies, furthermore, the aim was also to develop a system with a human-centric development approach (Asutay 2007).

According to Asutay (2007), the main ideologies of Islamic economics can be summarized as; having societal concern being a prerequisite, having socially God-conscious individuals who are concerned with social good who conduct economic activity in a way that is in accordance to Islamic constraints and having individuals who seek to maximize social welfare not just their own individual utility. Islamic economics is therefore a framework that can be used to solve human economic problems on the basis of the values, norms, laws and institutions derived from all religious knowledge sources in Islam (Haneef 2005). Given this, the Islamic financial system is based on the adherence to the Shari'ah meaning the prohibition of engaging in activities including riba (usury), gharar (uncertainty), maysir (gambling) and other activities deemed to be haram (unlawful) and prohibited under Shari'ah law (Ahmed 2015). These are considered to be the core tenants of Islamic finance.

Maqasid Al-Shari'ah

The objective of the Shari'ah according to scholar Al-Ghazali is to promote the well-being of people, which lies in preserving Maqasid Al-Shari'ah and ensuring the safeguarding of its five objectives in addition to serving public interest (Chapra 2008). Ibn Ashur, on the other hand, (2006) defined Maqasid from a broader dimension stating:

> The all-purpose principle of Islamic legislation is to preserve the social order or the community and insure its healthy progress by promoting the well-being and righteousness of the human being. The wellbeing and virtue of human beings consist of the soundness of their intellects and the righteousness of their deeds, as well as the goodness of the things of the world where they live that are put at their disposal.

Maqasid can be traditionally categorized under three levels; necessities (daruriyyat), needs (hajiyyat) and luxuries (tahsiniyyat) (Azman and Ali 2019). According to Dusuki and Bouheraoua (2011), Islamic scholar Abu Hamid Al-Ghazali defined maqasid emphasizing Shari'ahs concern with the safeguarding of the following five elements (also otherwise comprising the core elements of Maqasid Al-Shari'ah) as;

i. protection of faith (ad-din),
ii. protection of human self (nafs),
iii. protection of human intellect (aql),
iv. protection of future generations (nasl),
v. protection of resources/wealth (mal).

Moreover, Maqasid Al-Shari'ah has four main characteristics; firstly it is the basis of legislation serving the interests of all human beings saving them from harm, secondly it is universal in nature aiming to serve the interests of mankind and should be adhered to by all human beings, thirdly it is inclusive and fourthly, it is definitive and has not been derived from a singular text but rather a variety of texts (Dusuki and Bouheraoua 2011). Maqasid Al-Shari'ah can simply be defined as the purpose or objectives behind the Shari'ah (Islamic law) or Islamic ruling (Auda 2008); furthermore, it can also be understood as the goal of the Shari'ah, essential to the achievement of solving individual and societal problems (Zain and Ali 2017).

By understanding the above definitions, it can be understood that Maqasid Al-Shari'ah promotes cooperation and mutual support within a society; moreover, this manifests to the realization of "maslaha" (public interest) which Islamic scholars consider to be the objective and intention of the Shari'ah (Dusuki and Bouheraoua 2011). Islamic jurists define maslaha as a benefit or public interest in the literal sense and unrestricted public interest in the absence of regulation by God when no textual authority can be otherwise found in technical terms (Awang et al. 2014). Kamali (2008) expands the meaning of maslaha to the social justice realm and states that maslaha's aim is to achieve justice not only in the punitive sense but also in a harmonious sense, distributing justice and establishing an equilibrium of benefits and advantages in society.

The meaning of maslaha is therefore aligned with the Maqasid al-Shari'ah indirectly with the objective to not impose difficulties on people but rather to promote and facilitate welfare and fairness among them (Awang et al. 2014). Islamic finance therefore should aim to promote the general wellbeing of people (in other words maslaha) which is a manifestation of the Maqasid al-Shari'ah. Unlike in conventional economics where most decision-making theories such as the marginal productivity and utility theories are rooted in the goal of profit maximization, the five elements of the Maqasid Al-Shari'ah can be used as a multidimensional decision-making tool to assess the role of the firm in preserving and promoting public interest and the welfare of society, the economy and the family institution (Larbani and Mohammed 2011).

The aim of this chapter is to assess the role of traditional faith-based funding mechanisms, in particular Islamic finance, in humanitarian response therefore this chapter is organized as follows; the next section will include a literature review to give an overview of Islamic finance and its similarities to the socially responsible investing industry. Furthermore, the section will also analyze the various humanitarian uses of Islamic financial instruments such as sukuk, section two will also critically analyze the relevant literature on Islamic social finance and its key instruments. Section three will use the case study method to evaluate the International Federation of the Red Cross and the Red Crescent (IFRC) drought assistance program and in particular, analyze the use of Islamic social finance instruments and the lessons learned from the program. Section four will conclude the chapter, summarize the key findings and give recommendations for further research on this topic.

LITERATURE REVIEW

Socially Responsible Investing and Islamic Finance

Socially responsible investing and finance has emerged as a subsector within the financial industry and has been gaining traction in recent years. Socially Responsible Investing (SRI) is an investment form that considers social, ethical and environmental components alongside financial returns, in the investment decision-making process. Although the terms ethical investment and SRI are often used interchangeably, the main difference between the two is ethical investments describe an investment methodology that typically involves negative screening, while SRI, on the other hand, is an investment methodology that entails positive screening—in other words, SRI involves the process of searching for companies that uphold good practices and corporate behavior (Scanlan 2005). SRI therefore broadly encompasses various investment approaches from negative screening to complex corporate engagement, such as ethical investments, social investments and environmentally oriented investments, i.e., green and mission-oriented investments (Barom 2013).

Similar to SRI, Islamic finance derives its roots from religious doctrine and was established in adherence to Shari'ah principles such as equality and prosperity. In addition, Islamic finance has many principles that make it similar to SRI such as the emphasis on social justice and welfare (Biancone and Radwan 2019). Economic development and growth, in addition to social justice, are therefore the foundational components of the Islamic economic system whereby all societal members are given the same opportunities and "a level playing field" to advance themselves (Bennett and Iqbal 2013). Theoretical, philosophical and historical analysis of Islam reveals great importance placed on the social sector (Mohamad et al. 2016). Islamic finance exists to achieve multidimensional objectives such as religious, social and economic (Ahmed and Barikzai 2017), given this, Islamic finance requires not just the measurement of a bottom line but an approach whereby social impact is also taken into consideration and measured with financial returns (Mohamad et al. 2016).

Other similarities between SRI and Islamic finance includes screening which excludes investing in certain industries deemed immoral and viewed as having negative social impacts such as gambling, speculation, alcohol, tobacco, weapons manufacturing and pornography industries (Wilson 1997; Fowler and Hope 2007). Although both approaches are based on considering the social and ethical impacts that companies have (Charfeddine et al. 2016) as well as Islamic finance and investments being considered under the broader category of SRI—the practice of SRI and Islamic finance vary. For example, unlike SRI, Islamic finance imposes greater financial screening ensuring the level of conventional debt does not exceed the "Shari'ah tolerated threshold" given that interest-based activities are not deemed to be Shari'ah compliant (Bin Mahfouz and Hassan 2012). Furthermore, another distinction between SRI and Islamic finance is SRI places emphasis on issues such as environmental risk, corporate governance and the ethical practices of corporations and its engagement with its stakeholders both internal such as employees and external such as investors, customers and the wider society in which the corporation operates in.

Shari'ah compliant firms are those that incorporate Islamic values into their company and their core activities are expected to be in compliance with Shari'ah principles. This differs from the conventional investing strategy view which is to achieve the financial goal of maximized risk adjusted returns (Hussain et al. 2019). However, the notion of compliance is aligned with SRI where investors usually focus on ethical considerations in addition to enhancing the agenda of social welfare (Haji and Ghazali 2013; Haque et al. 2019). These types of firms appeal to investors who are seeking satisfactory financial returns for their investments in addition to wanting to consider social returns and ensure the social impact of their investments are aligned with their beliefs (Hussain et al. 2019). In such investments, investors are able to integrate their moral standards into their portfolios which act as a disciplining tool to regulate and manage corporate behavior (Ahmed 2010). Also, as both Islamic finance and SRI investments are based on similar guiding principles of corporate social responsibility this allows "harmonization of corporate objectives with the aim of achieving sustainable outcomes for all stakeholders from the social, economic and environmental perspective in the medium and long-term" (Franzoni and Allali 2018).

Sukuk

With the increase in poverty, rising unemployment rates, the emergence of several social issues, the restraints of public spending and severe environmental problems as well as other factors—there has been a call for an innovative approach to overcome such issues through the blending of social objectives with businesses, as the public sector and governments have been urged to find innovative and alternative sources of financing such as tapping Islamic financial instruments (Biancone and Radwan 2019). Considered to be a form of "ethical, inclusive, and socially responsible finance, because it connects the financial sector with the real economy and promotes risk sharing, partnership style financing, and social responsibility," Islamic financial instruments have emerged as effective tools to fund development globally (Asutay 2013). Sukuk, in particular, have been used to raise funding for humanitarian purposes.

Sukuk is the plural of Arabic word "sakk" which translates to mean certificates, according to the Accounting and Auditing Organization for Islamic Financial Institutions (AAOIFI), sukuk can be defined as "certificates of equal value representing undivided shares in ownership of tangible assets, usufruct and services or (in the ownership of) the assets of particular projects or special investment activity" (AAOIFI 2008). Sukuk are often referred to as "Shari'ah compliant bonds or Islamic bonds" and are fixed-income non-interest bearing securities. Moreover, they are certificates that represent a participation right to the underlying assets (Iqbal and Tsubota 2006).

Overall, there are at least fourteen types of sukuk structures recognized by AAOIFI based on assets, debt, equity and services (Ahmed 2010). The main three types of Sukuk structures are; Murabaha which are referred to as cost-plus sale contracts, Musharakah/Mudharabah which are contracts based on profit-sharing arrangements and Ijarah which are often referred to as sale-and-lease-back contracts (Jobst 2007). The first two sukuk ever issued were by Malaysian corporates in 1990 and 1996, and Malaysia was a key driving force in building the early sukuk market (Hussain et al. 2015). Furthermore, Malaysia is considered to be the center and one of the largest sukuk markets in the world, accounting for nearly one-third of all sukuk issued in 2014 (Smaoui and Khawaja 2017).

Sukuk are the most popular Islamic financial instruments and are a commonly used tool to raise capital by corporations and sovereigns in Muslim-majority countries as well as Western states. Sukuk have experienced tremendous growth and are frequently traded in Western conventional financial markets in addition to Shari'ah compliant markets demonstrating their mainstream viability and the globalization of Islamic finance. For over the past decade, there has been a surge in sukuk issues and the market has become the fastest growing sector of the Islamic financial services industry surpassing the Islamic banking sector on a growth basis (Smaoui and Khawaja 2017). Although sukuk have been responsible for much of the growth of the Islamic finance industry, despite their popularity and wide use, sukuk are often criticized. For instance, they are criticized for their lack of contribution to economic growth and development, their structure and whether they are truly Shari'ah complaint and the nature of the underlying assets as questions are raised as to whether investors have a claim to underlying assets, among other criticisms (Ahmed 2015).

Socially Responsible Investing (SRI) Sukuk

The inaugural SRI sukuk used for humanitarian purposes, in this case raising funds for vaccines, was issued by the World Bank and the International Finance Facility for Immunization (IffIm) in 2014 which was for the amount of USD$500 million. The issue was the first SRI sukuk of its kind that provided both financial returns (paying a competitive return) and social impact returns by supporting the immunization of children in developing countries (Bin Syed Azman and Ali 2016). The SRI sukuk issue was well received by investors (Badeeu et al. 2019) and the success of the transaction led the IffIm and the World Bank to issue yet another SRI sukuk in 2015 worth an estimated USD$200 million—also for vaccination purposes (Bennet 2015).

In 2015, Malaysian sovereign wealth fund Khazanah Nasional Berhad (Khazanah) launched the first SRI sukuk from its newly established SRI framework introduced by the Securities Commission of Malaysia in 2014. The sukuk was the first to be issued under a RM1 billion SRI sukuk program for a period of 25 years, with the first tranche issued worth to be RM100 million (Bin Syed Azman and Ali 2016). Funds raised from the sukuk were allocated to a Trust School program, a type of public–private partnership with the government that aimed to enhance the accessibility of quality education in Malaysia, and was managed by the not-for-profit organization Yayasan AMIR.

Both sukuk programs demonstrated how Islamic financial instruments such as sukuk can be used to fund social programs, yield financial returns as well as have a positive humanitarian impact by improving healthcare or the quality of education.

Islamic Social Finance

The traditions of charitable giving have long been established in Islam. Charitable models, practices and approaches have been collectively categorized under the umbrella term of "Islamic social finance" (ISF). Not only does ISF date back to the beginning of Islam but according to Olanrewaju et al. (2020), it "represents formidable social economic structures that redistribute wealth to reducing poverty in the society." As drawn from the provisions of the Shari'ah, social and economic justice, shared prosperity and inclusivity, are key tenets underlining ISF. In essence, it refers to the delivery of financial services to achieve socioeconomic welfare for societies most vulnerable members (Lawal and Ajayi 2019).

The teachings of Islam advocate for a comprehensive method to development and places great emphasis on social welfare, this holistic developmental approach has three dimensions; (1) the individual, (2) the environment (3) and society (Mirakhor and Askari 2010). Furthermore, Bennett and Iqbal (2013) highlight the central economic tenet in Islam is to "develop a prosperous, just and egalitarian economic and social structure in which all members of society irrespective of their beliefs and religious affiliations could maximize their intellectual capacity, preserve and promote their wealth, and actively contribute to the economic and social development of society."

From an Islamic perspective, the economy is comprised of two main activities; namely commercial and non-commercial. These two activities are not supposed to function in parallel, in fact, each type of activity is rooted in the other creating an economic equilibrium (Jouti 2019). Furthermore, according to Jouti (2019), this approach is used to build social finance ecosystems with for-profit and not-for-profit organizations working in tandem to develop social and economic welfare. Moreover, the ISF system represents the various Islamic institutions that aim to protect the social welfare, in addition to individual interest, to enhance economic activities and improve public wellbeing (Olanrewaju et al. 2020).

There are three types of institutions the ISF ecosystem is comprised of that are aligned with Shari'ah principles (Islamic Research and Training Institute 2014); the first type of institution is based on philanthropy and includes instruments such as Zakat, sadaqah and waqf; the second type is institutions based on mutual cooperation with instruments such as qard al-hassan (benevolent loan) and kafala (guarantee); and the third type is contemporary Islamic microfinance institutions.

Instruments of Islamic Social Finance

Although ISF instruments can be divided into three main categories (i.e., philanthropic-based instruments, cooperative-based instruments and other forms of Islamic financial services such as Islamic microfinance), for the purpose of this study, this section is limited to the discussion of the traditional instruments based on philanthropy, such as sadaqah, waqf, qard al-hasan and zakat. These instruments will be discussed in turn.

Sadaqah

Sadaqah is another form of almsgiving and a voluntary form of giving. In general, there are few restrictions on what charitable purposes sadaqah can be used for and it can be distributed throughout the year. Furthermore, the meaning of sadaqah is not only restricted to monetary terms and can include a wide array of charitable acts such as smiling at a stranger, saying kinds words, acts of compassion such as feeding the poor, passing on knowledge and so on. Therefore, given its discretionary purpose, it is often difficult to calculate the amount monetary sadaqah proceeds given at any one point. In practice, monetary forms of sadaqah can be given directly to the beneficiary rather being channeled through an intermediary or institution. From a humanitarian perspective, using sadaqah as a tool to finance humanitarian crises proves to be challenging as it is less structured, due to its irregularity and difficulty in calculation.

Waqf

According to AAOIFI, waqf (plural awqaf) is "making a property invulnerable to any disposition that leads to transfer of ownership and donating the usufruct of that property to the beneficiaries" (AAOIFI 2015). In general, waqf can be broadly defined as a type of Islamic endowment in the form of cash, real estate or any form of private wealth, which is donated for a charitable purpose in perpetuity and directed by the endower (Ahmed 2007). One of the main characteristics of waqf is the idea of doing charity out of goodness. According to Ahmed (2007), the objective of waqf "may be for the society at large, including the provision of religious services, socio-economic relief to the needy segment, the poor, education, environmental, scientific, and other purposes." Waqf is often described as one of the most important institutions having provided the foundation for Islamic civilization as it was intertwined with the social economy and the comprehensive religious life of Muslims (Hennigan 2004; Abbasi 2012). Public sector entities such as mosques, school and hospitals are often financed through awqaf (Abbasi 2012).

Qard Al-Hasan

As mentioned earlier, the prohibition of interest in lending transactions is one of the key tenets of Islamic finance. However, there are two types of loans (qard) that are permissible with one of them being a benevolent loan otherwise known as qard al-hassan. Qard al-hasan can be defined as a loan for members of the community who are under financial distress.

Some of the characteristics that differentiates qard al-hasan from conventional loans are (Iqbal and Shafiq 2015);

i. it is a non-interest-bearing loan whereby the borrower is morally obliged to repay the principal if they can and depending on their financial capacity to do so;
ii. the aim of qard al-hasan is to crowd in those who are poor with limited access to finance to help them become part of the economic system;

iii. the reward for lenders to extend credit in the form of qard al-hasan is for benevolent and spiritual purposes, and this element supersedes any monetary reward.

Zakat

While there are many definitions of what constitutes as poverty, according to Hassan (2010), Islam defines poverty as individual failure to meet the objectives of Maqasid Al-Shari'ah. Sadeq (1997) suggests that from an Islamic perspective, poverty can be eliminated by positive measures, preventative measures and corrective measures. Of the three set of measures, payment of zakat is classified as a corrective measure of poverty eradication. Zakat can therefore be seen as an instrument with an underlying humanitarian impact that enhances human welfare when conceptualized within the general objectives of Maqasid Al-Shari'ah (Billah 2018).

The literal meaning of zakat is to grow or increase and the word is mentioned frequently in the Holy Quran (Qaradawi 1999) and Zakat is also loosely translated to mean tithes or almsgiving (Al Haq and Farooq 2017). It stems from the belief that all wealth belongs to God and those with means have a moral obligation to assist those in need (Malik 2016). As one of the five pillars of Islam, Muslims are obligated to pay zakat on an annual basis to assist the poor and through the act of giving zakat, it is believed that wealth is purified (Bremer 2015).

Zakat therefore is a mandatory form of almsgiving whereby Muslims donate 2.5% of savings which is payable by Muslims whose wealth exceeds a minimum threshold known as the nisab (Ali and Hatta 2014). The nisab as a threshold is generally low so that a larger group of people are obligated to pay zakat and redistribute wealth (Pollard et al. 2015). Moreover, at the end of the holy month of Ramadan, zakat al-fitr is an obligatory donation that must be paid, it equates to the cost of feeding one person (Atia 2011).

Zakat has a number of social and economic values, from a social perspective it is often seen as means to redistribute wealth, alleviate poverty, eliminate greed and enhance the overall well-being of society (Gambling and Karim 1986; Sulaiman 2003). The main objective of zakat is to relieve poverty and redistribute wealth to promote harmony within the Muslim community (Atia 2011). Additionally, zakat can also be seen as a mechanism to encourage "socially oriented behavior" and discourage greediness (Nadzri et al. 2012). Historically, zakat has played a crucial role in alleviating poverty, and it has been documented that the practice of zakat has been used since the early Muslim migrants in Mecca with the objective of helping the poor and needy (Qaradawi 1999). Zakat is payable on business revenues, assets, gold, silver and savings at a rate of 2.5% (Gambling and Karim 1986; Hamid et al. 1993; Mohamed 2007; Mohamed Ibrahim 2001; Lewis 2001; White 2004). According to Gambling and Karim (1986), unlike conventional taxation, zakat is viewed as a means of purifying wealth rather than just a religious obligation.

There are certain restrictions on how Zakat is collected and distributed as it has to be in a manner that is harmonious with the Shari'ah. Furthermore, Zakat can only be distributed to people who fall under specific categories (asnafs) and in total, there are eight categories of eligible zakat recipients that include (Nasir and Zainol 2007);

    i. the poor,
    ii. the needy,
    iii. the wayfarer,
    iv. the heavily indebted,
    v. new converts to Islam,
    vi. freeing those enslaved,
    vii. and spending in the cause of God.

Zakat is a significant source of income for Muslim humanitarian organizations operating around the world and in recent years, there has been an increased interest in mobilizing Zakat for poverty alleviation in Muslim majority countries and for humanitarian purposes. In the Muslim world, there has been a renewed interest in zakat by conservatives as well as reformers who perceive zakat to be a tool for attaining social justice (Fauzia 2013). According to Shaikh (2016), the value of Zakat is estimated to be around US$187 million within countries in the Organization of Islamic Co-operation (OIC), with other estimates forecasting higher. The global annual value of zakat is estimated to be between US$200 billion to US$1 trillion; however, the potential of zakat has not yet been realized as informal giving of zakat is larger than formal donations made through Islamic organizations or institutions (Biancone and Radwan 2019).

Traditionally, Zakat should be spent within one lunar year in the area in which it was collected (Stirk 2015); however, it could be distributed to areas of greater need. For example, in Kuwait, the Shari'ah Board of the official Zakat House permits zakat funds to be transferred to areas which are more in need (Ahmed 2004), and in Indonesia, zakat collected in a village does not have to be used solely in that village and can be transferred to other villages with greater needs (Fauzia 2013). Zain and Ali (2017) suggest that for countries where there are surplus zakat funds, they should channel zakat to other countries which are in greater need and to consider those funds as a type of progressive investment from one country to another—especially given that the key objective of zakat is to alleviate poverty and redistribute wealth within the Muslim community globally to enhance shared prosperity.

The literature on zakat can broadly be categorized under studies conducted on the collection, distribution and management or administration of zakat, with subcategories stemming from these categories; however, there is a dearth of studies researching and examining the distribution of zakat in terms of what is distributed, how it is distributed, whom it is distributed to and the impact of zakat on improving the lives of recipients (Ahmed et al. 2017). Therefore, the rest of this chapter will focus on the potential of zakat as a funding mechanism and tool to enhance humanitarian impact.

Case Study: International Federation of the Red Cross and the Red Crescent (IFRC) Drought Resistance Program

This section will evaluate the use of Islamic social finance instruments in a humanitarian drought assistance program implemented by the International Federation of the Red Cross and the Red Crescent Societies (IFRC) in Kenya. This program was chosen for this study because it demonstrates the potential of ISF to alleviate poverty and provide humanitarian assistance. Moreover, the case also illustrates the international nature of zakat, as the program utilized zakat proceeds raised in Malaysia which was then channeled to Kenya.

The aim of this section is to assess the role and efficacy of traditional Islamic social finance mechanisms in the humanitarian sector. The purpose in doing so is to provide a more holistic understanding of the issues that arise when utilizing zakat funds in this manner. To give some context, a brief background will be given on climate-related disasters and their humanitarian impact before the analysis of the case is conducted.

Humanitarian Impact of Climate-Related Disasters

Climate-related disasters such as droughts create negative economic, environmental, social and security impacts, especially in fragile contexts (Muller 2014). Overtime, a confluence of factors, such as climate-related disasters, population growth, increased vulnerability and higher temperatures, will result in frequent food shortages and vector-borne diseases (IPCC 2007) and as droughts become more frequent, it could result in further food insecurity, famine, mass displacement, epidemics and conflicts. Furthermore, climate change is set to increase the frequency, intensity and duration of droughts further impacting food, water and energy in a disproportionate way according to the World Meteorological Organization (Muller 2014).

It is estimated that over a billion people live in arid and semi-arid lands (ASALs), of which Muslims constitute a significant proportion (Koohafkan and Stewart 2008). ASALs are highly exposed to the negative impacts of climate change and are characterized by extreme rainfall variability and frequent droughts, making the need to adapt to water shortages essential for smallholder farmers (Enfors and Gordon 2008; Stringer et al. 2009). In a severe drought, a lack of rainfall results in inadequate water supply for plants, animals and human beings and could last for several years, having a damaging effect on life and livelihoods (Muller 2014). Moreover, given that small-scale rain-fed agriculture is the main source of livelihood in the ASALs of sub-Saharan Africa, in times of drought, livelihoods are disproportionately impacted leading to greater socio-economic inequalities (Mavhura et al. 2015).

In the past three decades, a series of droughts have taken place in East Africa (Kallis 2008), and in the region, Somalia, Ethiopia and Kenya have been the countries most impacted by severe drought events (Meier et al. 2007). A surge of frequent drought events in the region date as far back as the early 1980s (Funk et al. 2013; Bayissa et al. 2015), and more recently, there has been an increase in below-average rainfall in the region (Nicholson 2014; Fenta et al. 2017).

From a humanitarian perspective, the most vulnerable people are usually most impacted by climate-related disasters as the compounding effects of small-scale weather events increasingly undermines people's abilities to manage and recover from disasters (Muller 2014). For instance, droughts in East Africa are a recurring phenomenon with significant humanitarian impacts (Anderson et al. 2012), during the 2008–2010 drought, over 13 million people were impacted in East Africa (Muller 2014) also, in the 2010–2011drought, a severe food crisis was triggered leading to widespread famine impacting an estimated 12 million people (ACTED 2011; AghaKouchak 2015). Furthermore, according to the Food and Agriculture Organization (FAO), 124.2 million people in East Africa were in recent years undernourished—the equivalent of one in nine people (FAO 2015).

Climate change and environmental factors are not only major triggers of conflicts but also exacerbate them, especially in pastoralist communities (Kevane and Gray 2008). According to Sachs (2006), drought-induced famine is more likely to trigger conflict in an area that is impoverished and lacking a cushion of financial or physical resources. The humanitarian impacts of climate-induced disasters can not only be devastating but coupled with the increase in the number of conflicts, has resulted in the widening of the humanitarian funding gap as a growing number of people are in need of humanitarian assistance, in 2016 alone the humanitarian financing gap was estimated to be US$15 billion (Georgieva et al. 2016).

IFRC Drought-Assistance Program

In early 2017, Kenya experienced one of the worst droughts in the country's history and at the peak of the drought, it was estimated 2.7 million people were in need of humanitarian assistance. The IFRC joined forces with the Kenya Red Cross and other aid groups to launch an international emergency appeal with the aim of raising funding to provide people with immediate access to water, food and healthcare for recovery efforts. Jointly, the IFRC, Kenyan Red Cross and other humanitarian organizations pioneered a drought-assistance program in southern Kenya. The aim of the program was twofold, first, to provide humanitarian relief to communities impacted by the drought occurring in the Kenyan county of Kitui with the objective of enhancing water security and second, to improve livelihoods by tackling cash crop problems (IFRC 2018). Recognizing the increasing difficulties in fundraising and the need to redress funding shortfalls, the IFRC approached the Zakat Council of the

Malaysian State of Perlis to fund the drought assistance program. In doing so, the IFRC could not only explore the potential of ISF but the organization could also diversify its funding sources to move beyond traditional donors.

There are many reasons as to why the drought assistance program was created. The IFRC had already in place a ISF initiative that explores opportunities to utilize ISF instruments (within its broader innovative finance portfolio) along with National Red Cross and Red Crescent Societies, in order to aid communities experiencing humanitarian crises and development challenges (IFRC 2018). As part of their ISF initiative, the IFRC partnered with the Zakat Council of the Malaysian State of Perlis to mobilize zakat funds collected in the state of Perlis to fund a pilot project in the Kenyan county of Kitui.

Zakat management has strong government support in Malaysia and is regulated by law; however, each Malaysian state is free to enact zakat laws and manage zakat proceeds on a state level (Razimi et al. 2016). Islamic law and customs such as the administration of Zakat is under the jurisdiction of individual states (Ahmed 2004; Hudayato and Tohirin 2010; Mahamod 2011; Saad and Abdullah 2014). Moreover, each state is governed by a Sultan who is advised by a "State Islamic Religious Council"—the council responsible for all matters related to religion including the administration and management of Zakat of all fourteen Malaysian states (Nadzri et al. 2012). Given that different states have their own religious council, there are differences in how the definition of zakat is interpreted and how it is implemented, therefore, the practical distribution of zakat differs from state to state. For example, the state of Perlis distributes zakat to five of the zakat categories, namely the poor, the needy, zakat administrators, new converts to Islam and spending in the cause of God (fisabilillah) and of those five categories, distributes more funds to the cause of God (Nadzri et al. 2012).

According to the IFRC2018zakat funds collected by the Zakat Council of Perlis have largely been used at the state level and the council wanted to extend their impact therefore they partnered with the IFRC on their pilot drought assistance program with the intention of creating greater impact. Therefore, the zakat council of Perlis, the Kenyan County of Kitui, the IFRC and the Kenya Red Cross entered into a partnership whereby the zakat council of Perlis pledged USD$1.2 million of zakat funding to the program and the IFRC and the Kenya Red Cross would facilitate and implement the program in Kitui. The program was intended to simultaneously enhance water security by supporting communities in Kitui access clean water in addition to enhancing early drought recovery efforts by supporting the livelihoods local farmers. The fundamental aim of the partnership and program was to utilize ISF principles and instruments, such as zakat, to move beyond short-term charity and humanitarian relief, toward sustained social and economic impact through inclusivity and risk-sharing.

In order to enhance water security, the zakat funds raised for the program were used to repair existing boreholes, pumps as well as installing new ones (up to 30), in order to provide the county with clean water as well as provide livelihood opportunities for the county residents by giving them the ability to sell and distribute water (IFRC 2018). Furthermore, the program used zakat proceeds to enhance food security and livelihoods for county residents. The IFRC and the Kenya Red Cross used zakat funds to purchase green gram seeds then distributed 175,000 worth of seeds to selected households (directly supplying 2 kg to each household selected) to help families grow crops and yield a larger harvest with the intention of providing sustainable livelihoods in the long-term through saleable cash crops (IFRC 2018).

Program Results

Approximately 1.2 million people were provided with better access to clean water, 30 borewells were reformed, 245,000 acres were planted in the county of Kitui, 20,000 metric tons of green grams were harvested with a total market value of USD$20 million and 175,000 families were provided with livelihood opportunities and water supply (IFRC 2018). Overall, the program provided a saleable cash crop that could deliver food, education and healthcare for the county residents to reduce their reliance and to help them make their communities more sustainable. According to the IFRC (2018), the "transformative impact of the project" falls into three main areas;

i. the innovative use of ISF in a different context;
ii. the design and implementation of a program that supports long-term sustainability as opposed to short-term humanitarian relief for a community;
iii. the effective use of forecast-based financing.

First, the main "transformative impact" of the program was applying ISF instruments such as zakat, for international humanitarian purposes contributing to drought and famine relief efforts and poverty alleviation (Mohamed-Saleem 2020). Farmers in Kitui agreed to repay the cost of the green gram seeds donated to them and pay it forward to residents in the neighboring county of Garissa, creating a virtuous cycle as Kitui farmers contributed to their local economies and eventually become donors themselves (IFRC 2018).

Second, the "transformative impact" of the program was designing and implementing a program with the intention of supporting long-term sustainability rather than short-term humanitarian relief. The program was designed to empower the residents of Kitui through advancing their livelihoods and enhancing their food and water security by supplying clean water and cash crop yields that could be monetized to provide income, food, health and education to the community.

Third, another vital impact of the program was the utilization of forecast-based financing, this is a system that uses science-based weather forecasts to anticipate disasters in risk-prone areas based on certain metrics, if those metrics are triggered then funds and resources are automatically disbursed—before the disaster occurs (Coughlan de Perez et al. 2015). As mentioned earlier, Kitui county is situated in an area that is arid and semi-arid and is therefore prone to frequent droughts as a result of poor management of water catchment areas, deforestation, unsuitable soil conservation measures and land degradation (ICHA 2020). Furthermore, its unique terrain means food security efforts are threatened by crop failure and lack of pastures for livestock. The drought assistance program innovatively used forecast-based financing by applying the use of data, technology and futures analysis to identity the district in Kenya more likely to be negatively impacted by a drought (which in this case was Kitui) and the crop most likely to grow and survive in those difficult conditions which was the green gram (IFRC 2018).

Lessons Learned

There are many lessons to be learned from the IFRC drought-assistance program.

First, this case demonstrates how the successful mobilization zakat funds collected in one country can fund a humanitarian program in another. According to Zain and Ali (2017), one of the suggested solutions for the implementation of ISF instruments in OIC member states is countries that have a surplus from their fund collections or proceeds of assets should channel those surplus funds to other countries in need. Furthermore, they argue that these funds be used as "progressive investments" with proper management and with the purpose of assisting less fortunate Muslims. This implies that rather than mobilizing zakat funds for short-term humanitarian relief, zakat should be used as a type of investment vehicle to support Muslims living in poverty and for developmental purposes. Currently, the OIC has 57 member states, 54 of which are Muslim-majority states.

Kenya is not a member state of the OIC but has a significant Muslim population. Previous studies conducted on the nexus between religion and poverty found that religion played a significant role in poverty status (Keister 2007, 2008) and in Kenya, it has been found that Muslims were more likely to experience chronic poverty than Christians (Mberu et al. 2014). Muslims were also 2.17 times more likely to suffer higher lived poverty than Christians (Odhiambo 2019).

Given the poverty disparity between Muslims and non-Muslims in Kenya, it can be said that the drought-assistance program could have been implemented in a Muslim-majority ASAL area in Kenya that is prone to frequent droughts such as an area in the northeast of the country. This does not contradict with the humanitarian principles of impartiality as there are significant humanitarian needs in those regions; therefore, in this instance, zakat could have been used as a mechanism for wealth and income distribution to guarantee a fair standard of living for Muslims in Kenya who suffer from higher poverty levels and fulfill the essence of zakat as a social institution to strengthen the Muslim Ummah (global Islamic community) (Kaslam 2009).

In general, the IFRC drought-assistance program is an interesting case study of how zakat does not contradict humanitarian principles of impartiality. The case also highlights one of the main challenges of zakat which is the selection of beneficiaries, to ensure that those most in need receive zakat, and while Islam has established the eight eligible categories of zakat beneficiaries, the selection is at the discretion of the Zakat Funds' administrators (Machado et al. 2018) and distributors.

Second, this case demonstrates the viability of zakat funds to benefit both Muslims and non-Muslims alike for humanitarian purposes. One of the challenges of zakat is there is no general consensus on the use zakat funds for non-Muslims and policies vary across jurisdictions and zakat collecting organizations (Stirk 2015). Given that the topic of whether zakat can be distributed to non-Muslims is an ongoing debate, if zakat is considered as a poverty alleviation tool exclusive for Muslims, then its future role in the humanitarian sector may be limited as it contradicts the humanitarian principle of impartiality, namely that "humanitarian action must be carried out on the basis of need alone, giving priority to the most urgent cases of distress and making no distinctions on the basis of nationality, race, gender, religious belief, class or political opinions" (UNOCHA 2010).

Although Muslims represent the largest minority religious group, Kenya is not a Muslim-majority country. According to the Berkley Center for Religion, Peace & World Affairs (2017), Muslims represent the largest minority religious group in Kenya comprising approximately 11% of the population with the majority living in coastal and northeastern regions. Furthermore, the dominant religion in the county of Kitui is Christianity. Whether this program can be replicated again in the future using a different zakat donor is yet to be seen. While the Quran has established the eight categories for those who are eligible to receive zakat, no prioritization or proportional distribution formula is outlined in the Quran (Qardawi 2009) and the selection is at the discretion of the Zakat fund administrators (Machado et al. 2018).

One of the reasons this program was able to mobilize zakat funds collected in the Malaysian state of Perlis is due to the fact that in Malaysia, it is permissible to distribute zakat funds to non-Muslim groups domestically (Hamat and Hanapi 2017); therefore, it can be assumed that the same principle applies internationally. Given that the issue of whether non-Muslims can be beneficiaries of zakat is a widely contested issue among Islamic scholars, going forward, mobilizing zakat funds from a Muslim-majority state is dependent on the local zakat laws and regulations of those jurisdictions which needs to be taken into consideration.

Third, another lesson that can be learnt from this case is it demonstrates a sustainable way that ISF instruments, such as zakat, can be used for humanitarian purposes to enhance overall social, environmental and economic development.

One of the biggest drivers of humanitarian crises is climate change and natural disasters. OIC member states with significant Muslim populations are struggling with high burdens of humanitarian crises and natural disasters, and have been impacted more than any other part of the world

(SESRIC 2017). For example, in 2011 where there were 302 natural disaster events recorded globally and the first famine to affect Somalia in 20 years took place which demonstrated the interdependence of climate-induced drought with famine, fragility, poverty and governance (Ferris and Petz 2011).

Climate-related factors such as an increasing number of droughts globally; rapid and unsustainable urbanization coupled with insufficient public health infrastructure and social protection; water, food and energy scarcity; shortage of arable land; and the loss of biodiversity—are factors that are igniting and intensifying conflicts as often vulnerable populations compete for limited resources for survival (Burkle et al. 2014). Water scarcity is representative of a major political, economic and human rights issue that is further driving vulnerability and conflict (Gelsdorf 2011), also, it is one of the compounding and cascading effects of resource scarcity conflicts as it is essential for the production of food, biofuel and alternative energy (Burkle et al. 2014). Moreover, one of the causes of acute water scarcity is the compound extremes in hot and dry areas that can reduce crop yields.

As mentioned earlier, the IFRC drought-assistance program aimed to enhance water security by improving access to water and improve food security by increasing the crop yields of green grams by utilizing zakat funds. Given a significant proportion of Muslims live in ASALs, the IFRC program could be replicated in other ASALs to use zakat as a tool to help enhance climate-resilient development in those regions, help increase water and food security, in addition to enhancing livelihoods.

Fourth, this case not only demonstrates the innovative use of ISF instruments but also highlights the rising trend of south-south giving. The transition of some countries in the Global South from aid recipients to economic powerhouses is creating a space for them to challenge the traditional landscape of assistance programs (Quadir 2013). As humanitarian challenges exponentially increase, organizations are seeking to evolve their funding strategies and diversify their donor funding sources as they cannot solely rely on traditional donors. This case illustrates that as the humanitarian landscape becomes more challenging and it becomes increasingly difficult for funding needs to be met, organizations are looking to the Global South to facilitate south-south giving using innovative financial instruments such as ISF as key entry points.

Future Considerations

Although the pilot IFRC drought-assistance program yielded some benefits, for a similar project to be replicated elsewhere, there are some considerations that need to be taken into account.

First, a number of constraining factors impact the effective collection, management and distribution of zakat funds. Effective zakat management in essence aims to achieve the best possible outcome in timely zakat collection and distribution as prescribed in Islamic teaching (Johari et al. 2015). To achieve the desired economic and social goal of zakat in any society, effective and efficient management of funds is key (Idris and Ayob 2001). Individuals, governments and other stakeholders each have a unique role to play in the distribution of zakat (Farouk et al. 2017), and the role zakat plays in providing support to the poor and most vulnerable sects of society is partially determined by the type of collection, organizational structure and political motives (Machado et al. 2018). Furthermore, the extent to which zakat can be used to fund humanitarian interventions varies not only in the way it is collected and administered but it also depends on the country's domestic conditions which needs to be considered.

Also, for the sustainability of zakat institutions in Muslim countries and to maximize efforts in collecting and distributing zakat productively, trust is an essential factor (Bin-Nashwan et al. 2020). If there is a lack of trust in zakat institutions in Muslim communities, initiatives will be fragmented and lack a coordinated or systematic approach as those eligible to pay zakat may distribute funds as they deem necessary rather than channel it through an institution. According to Ahmad (2008), the main reason zakat has not been as effective as it should be, is due to the nature of how zakat is distributed by Muslims which traditionally has been directly or personally, rather than being channeled through an institution. A lack of trust in formal zakat institutions therefore is a constricting factor to scale and mobilize funds in order to address humanitarian needs as funds collected will be minimal. In other words, the potential of zakat as a poverty alleviating tool will not be realized or scaled if the amount of zakat collected is small due to a lack of trust (Kahf 1989).

For member countries of the OIC, zakat governance can be categorized into Muslim countries that institutionalize zakat and those that do not (Bin-Nashwan et al. 2020); furthermore, this category can be split further into countries where zakat is legally enforced by statute and coun-tries where zakat is voluntary (Powell 2009). Unlike some OIC countries, Malaysia is a unique case as not only is giving zakat mandatory by law but some form of incentives are also granted to zakat payers in order to encourage people to pay zakat (Rahim and Kaswadi 2014; Saad and Haniffa 2014). Also, Malaysia is often seen as having a relatively progres-sive zakat system where since 1991, the use of private zakat collection agencies has helped the country raise zakat funds by 400% (Hasan 2015). As humanitarian organizations seek to diversify their funding sources and consider innovative financial instruments and other donors from the Global South, and in particular member states of the OIC, these factors need to be taken into consideration.

Also, this case does bring up some questions about the role of faith-based instruments in the localization debate as more secular non-Muslim humanitarian organizations start to consider ISF. Following the 2016 World Humanitarian Summit, localization emerged as a central issue on the international humanitarian agenda (Barakat and Milton, 2020) and over the past few years, there has been a call for greater local-ization in humanitarian policy debates. The international humanitarian system has long been critiqued as being top-down and driven by the Global North (Gingerich and Cohen 2015), centralized and bureaucratic (Spiegel 2017) and criticized for being slow and risk-averse (Healy and Tiller 2014). Roepstorff (2020) argues that the call for localization has sparked a debate on how to track and channel funding more directly to local humanitarian actors, strengthen local capacity, enhance partner-ship models and to better integrate local views and voices of the affected population into humanitarian response.

Current discussions that are focused on localization of humanitarian aid tend to be focused on renegotiating the technical terms of financing without addressing the structural power imbalances that exist. Barbelet (2019) states that the localization debate has to explore increasing invest-ment in building the capacity of local actors by improving partnerships and ensuring coordination between international and local responders. In this case, the IFRC has a unique status in terms of how the organization brings together local, national and humanitarian action within a single network (Zyck and Krebs 2015) where local actors can contribute and potentially alter the power dynamics.

Innovative ISF instruments, such as zakat, can also help address power imbalances and transform the traditional funding of aid by shifting to an investment culture rather than sticking to the status quo which is an unsustainable giving model. This is as the drought-assistance program aimed to support long-term sustainability rather than just provide short-term humanitarian assistance and zakat funds raised were used to invest in other Kenyan counties such as Garissa—looking beyond the program itself. This case therefore demonstrates that not only can instruments such as zakat provide humanitarian relief in the short and medium term, but also in the long-term, it could provide donors with an exit strategy as zakat recipients become zakat donors themselves.

CONCLUSION: REALIZING THE HUMANITARIAN POTENTIAL OF ISLAMIC SOCIAL FINANCE

In recent years, ISF has been increasingly recognized as a viable tool to enhance social and economic development. Moreover, the ISF sector has developed significantly in recent years and has demonstrated its humanitarian potential in both Muslim majority and non-Muslim majority developing countries. Although ISF instruments have existed since the inception of Islam and have historically played a crucial role in poverty alleviation and humanitarianism, its potential has not been realized. For instance, it is estimated that 40% of the population in Muslim-majority countries live in poverty and account for between 40 and 45% of the world's poor (Momin 2017). Given the main objective of ISF is to alleviate poverty and hardship through the mobilization and distribution of financial resources in addition to its social, economic and humanitarian benefits, it has immense untapped potential in not only providing much-needed humanitarian assistance but also transforming the humanitarian sector.

This chapter analyzed the use of ISF instruments such as zakat for humanitarian purposes by a program launched by the IFRC in Kenya. In summary, the drought-assistance program demonstrated that ISF instruments such as zakat are indeed viable for climate-related humanitarian programs, and in particular droughts. Therefore, as the number of climate-induced humanitarian crises is forecasted to rise and given that a significant number of people living in drought prone ASAL areas are Muslim, this program could be replicated elsewhere in a context similar to Kitui.

This study was limited in part due to a shortage of data availability in the ISF sector. Although the case study used was useful in being able to draw inferences and conclusions, this method is limited as this case alone will not be able to draw a complete picture of the relationship between ISF and humanitarianism. Therefore, there are several recommendations that can be made for further study in this prevailing sector.

First, although this chapter solely focused on the humanitarian impact of ISF instruments, there is scope to consider the role other Islamic financial instruments can play, such as humanitarian sukuk (as briefly mentioned in the literature review), in alleviating humanitarian crises. Second, while the effective use of forecast-based financing was deemed as one of the transformative impacts of the program (IFRC 2018), it was beyond the scope of this study to consider this. Therefore, it would be useful to scrutinize this case further to assess how forecast-based financing mechanisms were used in the program and to what effect.

REFERENCES

Abbasi, M. Z. (2012). The classical Islamic law of Waqf: A concise introduction. *Arab Law Quarterly*, 26(2), 121–153.

Accounting and Auditing Organization for Islamic Financial Institutions (AAOIFI). (2008). *Standard No. 17*. Manama: AAOIFI

Accounting and Auditing Organization for Islamic Financial Institutions (AAOIFI). (2015). *Shari'ah standards*. Manama: AAOIFI.

ACTED. (2011). *East Africa: Drought predictable and predicted: Agency for technical cooperation and development*. Paris, France.

AghaKouchak, A., 2015. A multivariate approach for persistence-based drought prediction: Application to the 2010–2011 East Africa drought. *Journal of Hydrology*, 526, 127–135.

Ahmad, J. (2008). Zakat Management in Indonesian and Zakat Global Synergy, Paper for International Zakat Executive Development Programme, in Malaysia, 15–26 December 2008.

Ahmed, H. (2004). *The role of Zakah and aqwaf in poverty alleviation* (Occasional Paper No. 8). Jeddah, Kingdom of Saudi Arabia: Islamic Development Bank Group.

Ahmed, H. (2007). Waqf-based microfinance: Realizing the social role of Islamic finance. *World Bank*, 6–7.

Ahmed, H. (2010). Islamic finance at a crossroads: The dominance of the asset-based sukuk. *Butterworth's Journal of International Banking and Financial Law*, 25(6), 366–367.

Ahmed, M. (2015). Islamic project finance: A case study of the East Cameron Project. *The Journal of Structured Finance*, 20(4), 120–135.

Ahmed, H. I., and Barikzai, S. A. K. (2016). Objectives of Islamic finance achieved by Islamic Banks. *Al-Azva*, 31(45), 45–65.

Ahmed, B. O., Johari, F. and Abdul Wahab, K. (2017). Identifying the poor and the needy among the beneficiaries of zakat: Need for a zakat-based poverty threshold in Nigeria. *International Journal of Social Economics*, 44(4), 446–458.

Al Haq, M. A., and Farooq, M. O. (2017). Zakat, persistence of poverty and structural incidental segmented approach: A survey of literature. *Journal of Islamic Financial Studies*, 3(1).

Ali, I., and Hatta, Z. A. (2014). Zakat as a poverty reduction mechanism among the Muslim community: Case study of Bangladesh, Malaysia, and Indonesia. *Asian Social Work and Policy Review*, 8(1), 59–70.

Azman, S. M. M. S., and Ali, E. R. A. E. (2019). Islamic social finance and the imperative for social impact measurement. *Al-Shajarah: Journal of the International Institute of Islamic Thought and Civilization (ISTAC)*.

Anderson, W. B., Zaitchik, B. F., Hain, C. R., Anderson, M. C., Yilmaz, M. T., Mecikalski, J., and Schultz, L. (2012). Towards an integrated soil moisture drought monitor for East Africa. Hydrology and Earth System Sciences, 9, 4587–4631.

Asutay, M. (2007). Conceptualisation of second best solution in overcoming the social failure of Islamic banking and finance: Examining the overpowering of homoislamicus by homoeconomicus. *IIUM Journal of Economics and Management*, 15(2), 167–195.

Atia, M. (2011). Islamic approaches to development: A case study of Zakat, Sadaqa and Qurd al Hassan in contemporary Egypt. In *8th International conference on Islamic economics and finance*. Doha: Centre for Islamic Economics and Finance, Qatar Faculty of Islamic Studies, Qatar Foundation.

Auda, J. (2008). *Maqasid al-Shariah as philosophy of Islamic law: A systems approach*. International Institute of Islamic Thought (IIIT).

Awang, M. D., Asutay, M., and Azman Jusoh, M. K. (2014). Understanding of Maslaha and Maqasid al-Shariah concepts on Islamic banking operations in Malaysia. In *International conference of global islamic studies* (pp. 12–32).

Badeeu, F. N., Nafiz, A. R., and Muneeza, A. (2019). Developing regional healthcare facilities in Maldives through mudharabah perpetual sukuk. *International Journal of Management and Applied Research*, 6(2), 81–96.

Barakat S., and Milton S. (2020). Localisation across the humanitarian-development-peace nexus. *Journal of Peacebuilding & Development*, 15(2), 147–163.

Barbelet, V. (2019). *Rethinking capacity and complementarity for a more local humanitarian action*. Overseas Development Institute.

Barom, M. N. (2013). Conceptualizing a strategic framework of social responsibility in Islamic economics. *International Journal of Economics, Management and Accounting*, 21(1).

Bayissa, Y. A., Moges, S. A., Xuan, Y., Van Andel, S. J., Maskey, S., Solomatine, D. P., Griensven, A. V., and Tadesse, T. (2015). Spatio-temporal assessment of meteorological drought under the influence of varying record length: The case of Upper Blue Nile Basin, Ethiopia. *Hydrological Sciences Journal*, 60(11), 1927–1942.

Bennet, M. (2015). *Vaccine sukuks: Islamic securities deliver economic and social returns*. World Bank.

Bennett, M. S., and Iqbal, Z. (2013). How socially responsible investing can help bridge the gap between Islamic and conventional financial markets. *International Journal of Islamic and Middle Eastern Finance and Management*.

Berkley Center for Religion, Peace & World Affairs. (2017). *Faith and development in Focus: Kenya*. Retrieved from https://jliflc.com/wp-content/upl oads/2017/03/Faith-and-Development-in-Focus-Kenya.pdf.

Biancone, P. P., and Radwan, M. (2019). Social finance and financing social enterprises: An Islamic finance prospective. *European Journal of Islamic Finance*, 0(0).

Billah, M. M. (2018). Zakat: Its micro-entrepreneurship model and socio-humanitarian impact. In V. Cattelan (Ed.), *Islamic social finance entrepreneurship, cooperation and the sharing economy*. London, United Kingdom: Routledge.

Bin Mahfouz, S., and Hassan, M. K. (2012). A comparative study between the investment characteristics of Islamic and conventional equity mutual funds in Saudi Arabia. *The Journal of Investing*, 21(4), 128–143.

Bin-Nashwan, S. A., Abdul-Jabbar, H., Abdul Aziz, S. and Ismail, S. S. H. (2020). Challenges of zakah management in muslim developing countries. *International Journal of Zakat and Islamic Philanthropy*, 2(1), 22–31.

Bin Syed Azman, S. M. M., and Ali, E. R. A. E. (2016). Potential role of social impact bond and socially responsible investment sukuk as financial tools that can help address issues of poverty and socio-economic insecurity. *Intellectual Discourse*, 24.

Bremer, J. (2015). Zakat and economic justice: Emerging international models and their relevance for Egypt. In Third Annual Conference on Arab Philanthropy and Civic Engagement.

Burkle, F. M., Martone, G., and Greenough, P. G. (2014). The changing face of humanitarian crises. *The Brown Journal of World Affairs*, 20(2), 19–36.

Chapra, M. U., Khan, S., and Al Shaikh-Ali, A. (2008). *The Islamic vision of development in the light of maqasid al-Shariah* (Vol. 15).

Charfeddine, L., Najah, A., and Teulon, F. (2016). Socially responsible investing and Islamic funds: New perspectives for portfolio allocation. *Research in International Business and Finance*, 36, 351–361.

Coughlan de Perez, E., van den Hurk, B. J. J. M., Van Aalst, M. K., Jongman, B., Klose, T., and Suarez, P., (2015). Forecast-based financing: An approach for catalyzing humanitarian action based on extreme weather and climate forecasts. *Natural Hazards and Earth System Sciences*. 15(4), 895–904.

Dusuki, A. W., and Bouheraoua, S. (2011). The framework of Maqasid al-Shari'ah and its implication for Islamic finance. *ICR Journal*, 2(2), 316–336.

Enfors, E. I., and Gordon L. J. (2008). Dealing with drought: The challenge of using water system technologies to break dryland poverty traps. *Global Environmental Change*, 18(4), 607–616.

Food and Agriculture Organization of the United Nations (FAO). (2015). The state of food insecurity in the world 2015. In *Meeting the 2015 international hunger targets: Taking stock of uneven progress*. FAO, IFAD and WFP, Rome, Italy.

Farouk, A. U., Idris, K. M., and Saad, R. A. (2017). The challenges of zakat management: A case of Kano State, Nigeria. *Asian Journal of Multidisciplinary Studies*, 5(7), 142–147.

Fauzia, A. (2013). *Faith and the state: A history of Islamic philanthropy in Indonesia*. Leiden: Brill.

Fenta, A. A., Yasuda, H., Shimizu, K., Haregeweyn, N., Kawai, T., Sultan, D., Ebabu, K., and Belay, A.S. (2017). Spatial distribution and temporal trends of rainfall and erosivity in the Eastern Africa region. *Hydrological Processes*, 31(25), 4555–4567.

Ferris, E., and Petz, D. (2011). *A year of living dangerously: A review of disasters in 2010*. London: The Brookings Institution – London School of Economics.

Funk, C., Husak, G., Michaelsen, J., Shukla, S., Hoell, A., Lyon, B., Hoerling, M. P., Liebmann, B., Zhang, T., Verdin, J., and Galu, G. (2013). Attribution of 2012 and 2003–2012 rainfall deficits in Eastern Kenya and southern Somalia. *Bulletin of the American Meteorological Society*, 94(9), 45–48.

Fowler, S. J., and Hope, C. (2007). A critical review of sustainable business indices and their impact. *Journal of Business Ethics*, 76(3), 243–252.

Franzoni, S., and Ait Allali, A. (2018). Principles of Islamic finance and principles of corporate social responsibility: What convergence? *Sustainability*, 10(3), 637.

Gambling, T. E., and Karim, R. A. A. (1986). Islamic and 'social accounting'. *Journal of Business Finance and Accounting*, 13(1), 39–50.

Gelsdorf, K. (2011). Global challenges and their impact on international humanitarian action. Office for the Coordination of Humanitarian Affairs (OCHA).

Georgieva, K., Shah, N., Ibrahim, H., Jafar, B., Macnee, W., Manuel, T., Mohohlo, L., Sriskandarajah, D., and Wallstrom, M. (2016). Too important to fail—Addressing the humanitarian financing gap. High-Level Panel on Humanitarian Financing Report to the Secretary-General.

Gingerich, T. R., and Cohen, M. J. (2015). *Turning the humanitarian system on its head: Saving lives and livelihoods by strengthening local capacity and shifting leadership to local actors*. Oxfam America.

Haji, A. A., and Ghazali, N. A. M. (2013). The quality and determinants of voluntary disclosures in annual reports of Shari'ah compliant companies in Malaysia. *Humanomics*.

Hamat, Z., and Hanapi, M. S. (2017). The zakat fund and non-muslims in Malaysia. *International Journal of Academic Research in Business and Social Sciences*, 7(5), 494–505.

Hamid, S., Craig, R., and Clarke, F. (1993). Religion: A confounding cultural element in the international harmonization of accounting? *ABACUS*, 29(2), 131–148.

Haneef, M. A. (2005). Can there be an economics based on religion? The case of Islamic economics. *Post-Autistic Economics Review*, 34(3).

Hasan, S. (ed.). 2015. *Human security and philanthropy: Islamic perspectives and muslim majority country practices*. New York, New York: Springer.

Hassan, M. K. (2010). An integrated poverty alleviation model, combining Zakat, Waqf and microfinance. Seventh International Conference – The Tawhidi Epistemology: Zakat and Waqf Economy, Bangi.

Haque, A. U., Kot, S., and Imran, M. (2019). The moderating role of environmental disaster in relation to microfinance's non-financial services and women's micro-enterprise sustainability. *Journal of Security and Sustainability Issues*, 8(3), 355–373.

Healy, S., and Tiller, S. (2014). *Where is everyone? Responding to emergencies in the most difficult places*. London: Médecins Sans Frontières.

Hennigan, P. (2004). *The birth of a legal institution: The formation of the waqf in third-century A.H. Hanafi legal discourse*. Leiden: Brill.

Hudayati, A., and Tohirin, A. (2010). Management of zakah: Centralised vs decentralised approach. Paper presented at Seventh International Conference – The Tawhidi Epistemology: Zakat and Waqf Economy, Bangi Selangor.

Hussain, H. I., Grabara, J., Razimi, M. S. A., and Sharif, S. P. (2019). Sustainability of leverage levels in response to shocks in equity prices: Islamic finance as a socially responsible investment. *Sustainability*, 11(12), 3260.

Hussain, M., Shahmoradi, A., and Turk, R. (2015). An overview of Islamic finance. *Journal of International Commerce, Economics and Policy*, 7(01), 1650003.

Ibn Ashur, M. T. (2006). Treatise on Maqasid al-Sharia. Translated and anno-
tated Muhammad El-Tahir Al-Misawi (London and Washington: International
Institute of Islamic Thought), 87.

Idris, K. M., and Ayob, A. M. (2001). Attitude towards zakah on employment
income: Comparing outcomes between single score and multidimensional
scores. *Malaysian Management Journal*, 5(1&2), 47–63.

International Centre for Humanitarian Affairs (ICHA). (2020). *Integrating
forecast based action in an existing early warning system: Learning the
context*. Retrieved from https://www.forecast-based-financing.org/wp-con
tent/uploads/2020/04/Integrating-Forecase-based-Action.pdf.

International Federation of Red Cross and Red Crescent Societies (IFRC).
(2018). *The transformative power of international zakat: How zakat
support from Malaysia helped communities in Kenya recover from drought*.
Retrieved from: https://media.ifrc.org/wp-content/uploads/sites/9/2018/
05/Kenya_case_study_2018_DRAFT_002.pdf.

IPCC. (2007). Summary for policymakers. In S. Solomon, D. Qin, M. Manning,
Z. Chen, M. Marquis, K. B. Averyt, M. Tignor, H. L. Miller, H.L. (Eds.),
Climate change 2007: The physical science basis. Contribution of Working
Group I to the Fourth Assessment Report of the Intergovernmental Panel
on Climate Change. Cambridge University Press, Cambridge, England.

Islamic Research and Training Institute (IRTI). (2014). *Islamic social finance
report 2014*. Retrieved from: https://irti.org/product/islamic-social-finance-
report-2014/.

Iqbal, Z., and Tsubota, H. (2006). *Emerging Islamic capital markets: A
quickening pace and new potential* (No. 120140, 1–8). The World Bank.

Iqbal, Z., and Shafiq, B. (2015). Islamic finance and the role of Qard-al-Hassan
(Benevolent Loans) in enhancing inclusion: A case study of AKHUWAT.
ACRN Oxford Journal of Finance and Risk Perspectives, 4(4), 23–40.

Jobst, A. A. (2007). The economics of Islamic finance and securitization. *The
Journal of Structured Finance*, 13(1), 6–27.

Johari, F., Ali, A. F. M., and Aziz, M. R. A. (2015). A review of literatures on
current zakat issues: An analysis between 2003–2013. *International Review of
Research in Emerging Markets and the Global Economy*, 1(2), 336–363.

Jouti, A. T. (2019). An integrated approach for building sustainable Islamic social
finance ecosystems. *ISRA International Journal of Islamic Finance*.

Kahf, M. (1989). Zakat: Unresolved issues in the contemporary Fiqh. *Journal of
Islamic Economics*, 2(1), 1–22.

Kallis, G. (2008). Droughts. *Annual Review of Environment and Resources*, 33,
85–118.

Kamali, M. H. (2008). *Maqāṣid Al-Sharīãh made simple* (Vol. 13). International
Institute of Islamic Thought (IIIT).

Kaslam, S. (2009). Governing Zakat as a social institution: The Malaysian perspective. *Social and Management Research Journal*, 6(1), 15–32.

Keister, L. A. (2007). Upward wealth mobility: Exploring the roman catholic advantage. *Social Forces*, 85(3), 1195–1225.

Keister, L. A. (2008). Conservative protestants and wealth: How religion perpetuates asset poverty. *American Journal of Sociology*, 113(5), 1237–1271.

Kevane, M. and Gray, L. (2008). Darfur: Rainfall and conflict. *Environmental Research Letters*, 3(3), 034006.

Koohafkan, P., and Stewart, B. A. (2008). Water and cereals in drylands. London and Sterling, VA: Food and Agriculture Organization (FAO) & Earthscan.

Kuran, T. (1986). The economic system in contemporary Islamic thought: Interpretation and assessment. *International Journal of Middle East Studies*, 18(2), 135–164.

Larbani, M., and Mohammed, M. O. (2011). *Decision making tools for resource allocation based on Maqāṣid Al-Sharīʿah. Islamic Economic Studies*, 19(2), 51–68.

Lawal, I. M., and Ajayi, J. M. (2019). The role of Islamic social finance towards alleviating the humanitarian crisis in North-East Nigeria. *Jurnal Perspektif Pembiayaan dan Pembangunan Daerah*, 6(5), 545–558.

Lewis, M. K. (2001). Islam and accounting. *Accounting Forum*, 25(2), 103–127.

Machado, A. C., Bilo, C., and Helmy, I. (2018). The role of zakat in the provision of social protection: A comparison between Jordan, Palestine and Sudan (No. 168). Working Paper.

Mahamud, L. H. (2011). *Alleviation of rural poverty in Malaysia: The role of zakat: A case study*. Unpublished doctoral dissertation. The University of Edinburgh. Retrieved from: http://www.era.lib.ed.ac.uk/handle/1842/5554.

Malik, B. A. (2016). Philanthropy in practice: Role of zakat in the realization of justice and economic growth. *International Journal of Zakat*, 1(1), 64–77.

Mavhura, E., Manatsa, D., Mushore, T. (2015). Adaptation to drought in arid and semi-arid environments: Case of the Zambezi Valley, Zimbabwe. *Jamba: Journal of Disaster Risk Studies*, 1(7), 144–147.

Mberu, B. U., Ciera, J. M., Elungata, P., and Ezeh, A. C. (2014). Patterns and determinants of poverty transitions among poor urban households in Nairobi, Kenya. *African Development Review*, 26(1), 172–185.

Meier, P., Bond, D., and Bond, J. (2007). Environmental influences on pastoral conflict in the Horn of Africa. *Political Geography*, 26(6), 716–735.

Mirakhor, A., and Askari, H. (2010). *Islam and the path to human and economic development* (pp. 26–33). New York: Palgrave Macmillan.

Mohamed Ibrahim, S. H. (2001). *Islamic accounting—Accounting for the New Millennium?* Paper presented at the Asia Pacific Conference. Kelantan, Malaysia: International Islamic University Malaysia.

Mohammed, J. A. (2007). *Corporate social responsibility in Islam*. Unpublished doctoral thesis, Auckland University of Technology. Retrieved from Dissertations and Thesis Databases.

Mohamed-Saleem A. (2020). Localising humanitarianism, peace making, and diplomacy: The challenges facing muslim INGOs. *Journal of Peacebuilding & Development*, 15(2), 178–191.

Mohamad, S., Lehner, O. M., and Khorshid, A. (2015). A case for an Islamic social impact bond. Available at SSRN 2702507.

Momin, A. (2017). *Introduction to Sociology: An Islamic perspective*. New Delhi, India: Institute of Objective Studies.

Muller, J. C. Y. (2014). Adapting to climate change and addressing drought-learning from the red Cross Red Crescent experiences in the horn of Africa. *Weather and Climate Extremes*, 3, 31–36.

Nadzri, F. A. A., Rehman, R. A., and Omar, N. (2012). Zakat and poverty alleviation: Roles of Zakat institutions in Malaysia. *International Journal of Arts and Commerce*, 1(7): 61–72.

Nasir, N. M., and Zainol, A. (2007). Globalisation of financial reporting: An Islamic focus. In J. M. Godfrey, and K. Chalmers (Eds.), *Globalisation of accounting standards*. Cheltenham, United Kingdom: Edward Elgar.

Nicholson, S. E. (2014). A detailed look at the recent drought situation in the Greater Horn of Africa. *Journal of Arid Environments*, 103, 71–79.

Odhiambo, F. O. (2019). Assessing the predictors of lived poverty in Kenya: A secondary analysis of the Afrobarometer survey 2016. *Journal of Asian and African Studies*, 54(3), 452–464.

Olanrewaju, A. S., Shahbudin, A. S. M., and Zakariyah, H. (2020). A synthesis of the Islamic social finance for sustainable Islamic social enterprise: A four factor of production frame. *Journal of Critical Reviews*, 7(19), 9963–9974.

Pollard, J., Datta, K., James, A., and Akli, Q. (2015). Islamic charitable infrastructure and giving in East London: Everyday economic-development geographies in practice. *Journal of Economic Geography*, 16(4), 871–896.

Powell, R. (2009). Zakat: Drawing insights for legal theory and economic policy from Islamic jurisprudence. *Pittsburgh Tax Review*, 43(7), 43–101.

Qaradawi, Y. (1999). *Fiqh al zakah (Volume 1): A comparative study of zakah, regulations and philosophy in the light of Quran and Sunnah*. Saudi Arabia: King Abdul Aziz University.

Qardawi, Y. (2009). *A comparative study of Zakah, regulations and philosophy in the light of qur'an and sunnah*. *Fiqh Al Zakah*. Kingdom of Saudi Arabia: Centre for Research in Islamic Economics King Abdul Aziz University.

Quadir, F. (2013). Rising Donors and the New Narrative of 'South–South' Cooperation: What prospects for changing the landscape of development assistance programmes? *Third World Quarterly*, 34(2), 321–338.

Rahim, S., and Kaswadi, H. (2014). An economic research on zakat compliance among muslim's staff in UNIMAS. In *Proceeding of the International Conference on Masjid, Zakat and Waqf (IMAF 2014)* (pp. 53–64).

Razimi, M. S., Romle, A. R., & Erdris, M. F. (2016). Zakat management in Malaysia: A review. *American-Eurasian Journal of Scientific Research*, 11(6), 453–457.

Roepstorff, K. (2020). A call for critical reflection on the localisation agenda in humanitarian action. *Third World Quarterly*, 41(2), 284–301.

Saad, N. M., and Abdullah, N. (2014). Is zakat capable of alleviating poverty? An analysis on the distribution of zakat fund in Malaysia. *Journal of Islamic Economics, Banking and Finance*, 10(1), 69–95.

Saad, R. A., and Haniffa, R. (2014). Determinants of zakah (Islamic tax) compliance behavior. *Journal of Islamic Accounting and Business Research*, 5(2), 182–193.

Sachs, J. D. (2006). Ecology and political upheaval. *Scientific American*, 295, 37–37.

Sadeq, A. M. (1997). Poverty alleviation: An Islamic perspective. *Humanomics*, 13(3), 110–134.

Scanlan, C. (Ed.). (2005). *Socially responsible investment: A guide for pension schemes and charities*. Key Haven Publications.

Shaikh, S. A. (2016). Zakat collectible in OIC countries for poverty alleviation: A primer on empirical estimation. *International Journal of Zakat*, 1(1), 17–35.

Smaoui, H., and Khawaja, M. (2017). The determinants of Sukuk market development. *Emerging Markets Finance and Trade*, 53(7), 1501–1518.

Spiegel, P. (2017). The humanitarian system is not just broke, but broken: Recommendations for future humanitarian action. The Lancet.

Statistical, Economic and Social Research and Training Centre for Islamic Countries (SESRIC). (2017). *Humanitarian crises in OIC countries: Drivers, impacts, current challenges and potential remedies*. Retrieved from: https://www.sesric.org/files/article/573.pdf.

Stirk, C. (2015). *An act of faith: Humanitarian financing of zakat* (Briefing paper). Retrieved from: http://devinit.org/wpcontent/uploads/2015/03/ONLINE-Zakat_report_V9a-1.pdf.

Stringer, L. C., Dyer, J. C., Reed, M. S., Dougill, A. J., Twyman, C., and Mkwambisi, D. (2009). Adaptations to climate change, drought and desertification: Local insights to enhance policy in southern Africa. *Environmental Science and Policy*, 12(7), 748–765.

Sulaiman, M. (2003). The influence of riba and zakat on Islamic accounting. *Indonesian Management and Accounting Review*, 2(2), 149–167.

United Nations Office for the Coordination of Humanitarian Affairs (UN OCHA). (2010). *OCHA on message: Humanitarian principles*. Retrieved from: https://www.unocha.org/sites/dms/Documents/OOM-humanitarian principles_eng_June12.pdf.

Wilson, R. (1997). Islamic finance and ethical investment. *International Journal of Social Economics*, 24(11), 1325–1342.

White, L. S. (2004). *The influence of religion on the globalization of accounting standards*. Paper presented at the Christian Business Faculty Association Conference, organized by Abilene Christian University, San Antonio, TX. Abilene, Texas: Citeseer.

Zain, N. R. M., and Ali, E. R. A. E. (2017). An analysis on Islamic social finance for protection and preservation of Maqāṣid al- Sharīah. *Journal of Islamic Finance IIUM Institute of Islamic Banking and Finance*, 2117, 133–141.

Zyck, S., and Krebbs, H. (2015). Localising humanitarianism: Improving effectiveness through inclusive.

CHAPTER 6

Expanding Access to Financial Services

INTRODUCTION

Climate change is having a significant impact on economies globally. Previous empirical studies have established the direct impact of climate-induced disasters on economies (Cavallo et al. 2013; Felbermayr and Groschl 2014; Ferreira and Karali 2015; Mendelsohn et al. 2012; Alano and Lee 2016; Botzen et al. 2019) as with an increase in the frequency of natural disasters such as droughts, extreme temperatures, floods, landslides and storms (IPCC 2018)—climate change induced disasters are creating significant economic costs (Kling et al. 2021). However, although the impacts may differ across countries, according to Kling et al. (2021), there is a general consensus that the biggest impacts of climate change are being experienced in developing countries.

For example, in the Arid and Semi-Arid Lands of Africa, severe and prolonged droughts are devastating marginalized pastoral and agro-pastoral households on a continual basis (Wairimu Ng'ang'a and Crane 2020). In addition to chronic poverty, migration and land degradation, changing land tenure systems and increased human and livestock populations, these communities will be worst hit by the effects of climate change (Goldman and Riosmena 2013). For pastoralists in Eastern Africa, factors—such as the increasing challenges of mobility and significant reliance on climate-sensitive systems for their livelihood—are resulting

© The Author(s), under exclusive license to Springer Nature Switzerland AG 2021
M. Ahmed, *Innovative Humanitarian Financing*,
Palgrave Studies in Impact Finance,
https://doi.org/10.1007/978-3-030-83209-4_6

in acute food and nutrition insecurity as well as exacerbating regional inequality (Borgerhoff Mulder et al. 2010; Government of Kenya 2015).

The concept of building resilience to climate change has recently been at the forefront of development and adaptation conversations (Sudmeier-Rieux 2014) and this interest in resilience can partly be traced to the failure of development and humanitarian organizations to effectively anticipate as well as respond to the 2011 Horn of Africa Famine (Lautze et al. 2012; Haan et al. 2012; Majid and McDowell 2012; Hobbs et al. 2012; USAID 2012). The Horn of Africa Famine in addition to similar outcomes in the Sahel and South Sudan (Grist et al. 2014; Boyd et al. 2013), started conversations about "how to avoid such poor outcomes" according to Carr (2019).

There is a growing consensus that an effective way to enhance the climate resilience of marginalized populations is through the expansion of financial inclusion (Adegbite and Machethe 2020). According to Yoshino and Morgan (2017), financial inclusion can alleviate households' credit constraints and improve their financial stability. Furthermore, as an adaptation measure, financial inclusion can reduce poverty and enhance climate resilience as well as give small and medium enterprises better access to loans to enable them to invest in more profitable projects (Renzhi and Baek 2020).

The aim of this chapter is to explore to what extent enhancing financial inclusion, and in particular Islamic financial inclusion, through the expansion of access to financial services, can increase resilience to climate change. The remainder of this chapter is organized as follows; the next section will critically review the literature on microfinance and financial inclusion—both conventional and Islamic. Section three will evaluate the impact and effectiveness of the Program for Resilient Systems (PROGRESS) that was part of the wider Building Resilience and Adaptation to Climate Extremes and Disasters (BRACED) program implemented in Wajir, Kenya by Mercy Corps. Section four will conclude and provide recommendations for further study.

LITERATURE REVIEW

Microfinance

In order to understand the role of financial inclusion in developing countries, microfinance needs to be considered as microfinance not only acts as a vehicle for financial inclusion, but it also acts as a bridge for low-income and poor households and encourages more inclusion into the financial system (Adeola and Evans 2017). The Microcredit Summit (1997) defines microfinance as a program that extends small loans to impoverished people so they can engage in self-employed activities that generate income therefore allowing them to provide for themselves and their families. Modern microfinance emerged in the 1970s then developed into a global finance-development hybrid and was incorporated by the World Bank Structural Adjustment Programs (Mader 2018).

Microfinance has generally been considered as a poverty alleviating tool (Morduch 2000; Obaidullah 2008). According to Robinson (2001), microfinance is a reference to the provision of small-scale financial services to the poor. It has also been argued that microfinance has not only helped reduce poverty, but it has improved educational levels and has resulted in the creation and growth of millions of small businesses (Abdul Rahman 2007). Ahmed (2002) suggests that the expansion of microenterprises or small businesses can possibly result in the generation of employment, in addition to the development of poor countries. Microfinance is also increasingly becoming the core of financial inclusion (Morduch 2000).

For many developing countries, a lack of access to affordable financial services is considered one of the underlying issues for the level of poverty observed (Liñares-Zegarra and Wilson 2018) with advocates of microfinance suggesting that lack of access to formal financial services being a critical component as to why people in developing countries remain poor (Dahal and Fiala 2020). Proponents of microfinance claim that those who follow sound banking principles, such as microfinance, are those who will contribute to poverty reduction the most (Morduch 2000). In sub-Saharan Africa, the microfinance industry has been growing rapidly at a rate of over 10% per annum in the past two decades (Chikalipah 2018) and it has become an important instrument of consumption smoothing for the poor (Morduch 2000). Despite initial quasi-experimental studies conducted on the impact of microcredit showing huge potential in poverty reduction, particularly for female borrowers (Pitt and Khandker 1998), recent microfinance experimental studies show the impact of access to microcredit is not perhaps as transforming as it was initially thought to be, in other words the promise of microfinance in eradicating world poverty was not conclusive (de Oliveira Leite et al. 2020).

Microfinance Institutions (MFIs)

In the 1970s, there was an emergence of microfinance institutions (MFIs) providing credit to small entrepreneurs with the aim of enhancing economic growth and alleviating poverty (de Oliveira Leite et al. 2020). An MFI is a financial institution that aims to reduce poverty by providing finance to marginalized members of society (Berhane and Gardebroek 2011; Fianto et al. 2018). In addition to providing financial services to communities, MFIs also help to develop the business capacity of borrowers (Littlefield et al. 2003).

MFIs traditionally offer financial services, mostly loans, to individuals and groups in order to meet their short-term consumption and investment needs (Liñares-Zegarra and Wilson 2018), particularly to individuals and groups that have been traditionally excluded from the mainstream financial system (Cull et al. 2009; Hermes et al. 2011; Ledgerwood et al. 2013; Morduch 1999). According to Armendáriz and Morduch (2010), these institutions promised to do three things (1) increase access to the financial market, (2) improve gender equality and (3) strengthen communities.

MFIs generally do not solely focus on the bottom line as they aim to both deliver sound financial performance while also creating positive social impact. Moreover, MFIs commonly use innovative lending models based on joint liability targeting mostly female clients such as group lending and village banking (Baquero et al. 2018). In the absence of collateral and credit registries, this type of lending mechanism facilitates the ease of credit constraints more effectively and relies on mainly "soft information" and strong bank-borrower relations (Berger et al. 2005, 2011). The rapid expansion of MFIs led some of them to increase not only in size but also in complexity (de Oliveira Leite et al. 2020), this raised questions about the long-term sustainability of these types of institutions. Furthermore, MFIs were originally established as not-for-profit institutions that heavily relied on funding from donors and development agencies (Baquero et al. 2018); however, MFIs are increasingly trying to achieve self-sustainability and be less reliant on donations and subsidies and instead be more dependent on profits and commercial investors (Garmaise and Natividad 2010, 2013).

As such the "for-profit" MFI model has raised several questions in regards to ethics. Proponents of the model argue that the focus on sustainable operations and outreach are complementary while, on the other hand, opponents of the "for-profit" model argue that the model is inconsistent with the objective of MFIs as questions arise as to what extent can generating profits by lending to the poor be aligned with poverty alleviation goals (Baquero et al. 2018). Furthermore, several conventional MFIs have been accused of financial predatory behavior by not only charging excessive interest rates but also using aggressive collection methods (Boatright 2013). For over a decade, MFIs have been under increased scrutiny for creating over-indebtedness (Guérin et al. 2015), criticized for the high interest rates they charge (The Economist 2014), criticized for their focus on credit over other financial services (Mader 2015), scrutinized for the lack of impact on poverty reduction (Duvendack et al. 2011; Stewart et al. 2012) and their dubious record on women's empowerment (Fraser 2009; Karim 2011) and since then, financial inclusion has been introduced as a new term.

Financial Inclusion

The notion of financial inclusion is multidimensional and studies conducted offer various definitions as to what it means. In order to understand the concept of financial inclusion, it must firstly be understood and defined in the context of financial exclusion. The related literature on financial exclusion has defined it as part of a larger issue of the social exclusion of minority groups from the mainstream of society (Sarma 2016). Leyshon and Thrift (1995) provide one of the earliest definitions of financial exclusion and define it as the structural barriers preventing certain segments of society from accessing the formal financial system. Segrado (2005) argues exclusion from financial services can be as a result of high costs and prices, weak marketing, lack of financial literacy of products or self-exclusion in response to poor experiences and perceptions. Moreover, Sinclair (2001) defines financial exclusion as the inability to access financial services due to a range of reasons such as "problems with access, conditions, prices, marketing or self-exclusion in response to negative experiences or perceptions." And according to Carbo et al. (2005), financial exclusion is the lack of access of certain groups to the financial system similarly, Mohan (2006) defines financial exclusion as the inability of some societal groups to gain access to financial products and services. According to Hariharan and Marktanner (2012), the lack of financial inclusion is a multifaceted socioeconomic issue that is a result of a multitude of factors such as culture, religion, history, geography, inequality and economic structure and policy.

The above definitions indicate that financial exclusion occurs as a result of the systematic exclusion of minority groups operating at the fringes of society from the formal financial system. Taking the opposite of this perspective, financial inclusion can then be defined as the ease of access, availability and usage of financial services (i.e., savings, loans, investments, insurance, pensions) for all members of society (Sarma 2008). Financial inclusion is an all-encompassing definition and has many variations depending on the stages of development in different countries (Sharma and Kukreja 2013). Seman (2016) defines financial inclusion as a means of delivering financial services at an affordable cost to the less-privileged, low-income and more vulnerable members of the population. Additionally, Umar (2020) define financial inclusion as the ease of access to formal financial services such as credit, insurance, pension-saving products, savings and payments, at affordable costs. According to Kim (2016), financial inclusion is the accessibility of all adults to accounts, savings, loans, life insurance and general insurance products provided by formal financial institutions. Ambarkhane et al. (2016) extend this definition and also includes other non-banking financial services such as pensions, social welfare schemes and financial literacy.

Financial inclusion is not only viewed as only providing access to financial services (Beck et al. 2007a; Allen et al. 2012) but rather extends to continuous financial usage as access alone will not necessarily lead individuals and small enterprises to benefit off of formal financial services (Cihak et al. 2016). Having access to finance is important for many reasons, for entrepreneurs, it means they can take on more risk, invest more and contribute to growth in a positive way (Cumming et al. 2014). There have been many studies conducted in the fields economic development and poverty reduction that suggest the importance of financial inclusion (Zulkhibri 2016). For example, Beck et al. (2008, 2009) found empirical evidence that improved access to finance not only enhances growth but it also reduces income inequality and poverty.

Moreover, there are numerous benefits of financial inclusion. For instance, financial inclusion reduces income inequality (Adeola and Evans 2017), increases funding availability for "efficient intermediation and allocation" (Shankar 2013) and eases access to credit/loans, the ability to save for retirement, cope with shocks and start and grow businesses

(Demirgüç-Kunt et al. 2015). On the other hand, exclusive financial systems limit individuals to rely on informal financing hence limiting growth and exacerbating inequality (Beck et al. 2000; Klapper et al. 2006). While inclusive financial systems encourage investment in productive activities such as education and entrepreneurship, therefore, it is more likely to benefit poor and disadvantaged groups (Honohan 2008).

A number of theoretical models used have shown that a lack of access to finance can result in poverty traps and inequality (Aghion and Bolton 1997; Banerjee and Newman 1993; Galor and Zeira 1993) and studies have also shown that financial inclusion allows minority groups to build up capital to combat the predisposition to poverty (Agyemang-Badu et al. 2018). The literature on financial inclusion also indicates access to a bank account is not only a significant measurement for promoting financial capability but also to health as through enhancing financial stability, stress is reduced and health is improved (Ajefu et al. 2020).

The pursuit of financial inclusion to incorporate the "unbanked" population into the formal financial system has been a top priority for policymakers around the world. The topic has received increased attention in recent times, from both academia and policymakers, as inclusive growth has become an important goal for sustainable development (Demirguc-Kunt et al. 2018). Policymakers in several developing countries have implemented a number of policies to improve financial inclusion with the belief that more inclusive financial systems benefit economic development in several ways (Huang and Zhang 2020) such as through economic growth (Sarma and Pais 2011), financial efficiency (Beck et al. 2007b), financial stability (Morgan and Pontines 2014) and social welfare (Demirgüç-Kunt and Levine 2009).

One of the reasons expanding access to financial services for poor and marginalized groups has become a global policy priority is the need to increase economic growth and reduce poverty (Xu 2020) as a strong link between financial access to banking services and economic development and growth, has been established in the past decade by several studies (Honohan 2004, Demirguc and Klapper 2012a, among others). An increased interest in financial inclusion is also partly due to the realization that those with a lack of access to finance are disadvantaged when it comes to accumulating savings, investing and building assets to protect against risks and it negatively impacts economic growth and poverty alleviation efforts (Neaime and Gaysset 2018).

Islamic Microfinance

Conventional MFIs that attempt to enter into regions with a substantial Muslim population find it difficult as they are incompatible with Islamic law and principles (Karim et al. 2008). According to Rahman et al. (2015), 25–40% of Muslims surveyed decided to stay away from interest-based microfinance products from fear that they would be violating their religious beliefs similarly, surveys conducted in Jordan, Algeria and Syria by Karim et al. (2008), revealed that 20 to 40% of respondents cited religious concerns as reasons for not accessing conventional microfinance. This leads to the emergence of Islamic microfinance, a confluence of the rapidly growing microfinance and Islamic finance industries, as a new market niche (Abdul Rahman et al. 2015). According to Hassan et al. (2013), Islamic microfinance can be defined as: "[T]he process of providing small-scale financial services, based on Shari'ah (Islamic) concepts, to the poor who may be excluded from formal financial services, without putting any burden on the parties either in the form of interest or undue benefits." According to Weill (2020), the four main principles of Islamic microfinance are:

1. The prohibition of interest.
2. Lenders are rewarded through profit sharing.
3. Islamic MFIs cannot fund activities that are not compliant with Islamic principles such as gambling (maysir) (Chong and Liu 2009), alcohol or financially deal with MFIs that charge interest.
4. As uncertainty (gharar) is prohibited, contractual terms need to be clear and there should be no contractual uncertainty.

The emergence of Islamic microfinance as a new market niche offered microfinance products that were aligned to Islamic principles and therefore was able to offer an alternative path to poor Muslims who were not being served by conventional microfinance (Karim and Khaled 2011). Furthermore, the high demand for loans that were compliant with Islamic principles highlighted the need to provide religiously compatible products to underserved poor Muslims populations hence the creation of Islamic microfinance as a new market niche (Karim et al. 2008). Also, as a financial outreach tool, Islamic microfinance garnered a lot of interest given a large number of the world's poor live in Muslim-majority countries (Ahmad et al. 2020).

Similar to conventional microfinance, Islamic microfinance was created to meet the financial needs of the poor and those that were financially excluded, however, from an operational perspective, there's a considerable difference between Islamic microfinance and conventional microfinance. For example, some of the key distinctions between conventional MFIs and Islamic MFIs are their sources of funding, their investment and product portfolios and their management (Ahmad et al. 2020). Moreover, one of the biggest differences between conventional microfinance and Islamic microfinance is, in Islamic microfinance, the charging of riba (usury or interest) on microloans is prohibited (Ahmed 2002). Researchers have argued that the prohibition of riba is largely due to the "negative distributive justice and equity effects" of interest (Alkhan 2016; Khan 1986; Visser and McIntosh 1998) with the argument being that a riba-based system enriches the rich while impoverishing the poor (Alkhan and Hassan 2020).

Islamic Microfinancial Institutions (Islamic MFIs)

Similar to conventional MFIs, Islamic MFIs are institutions that provide financial services to those that are unable or experience difficulty in gaining access to formal financial institutions. However, Islamic MFIs adhere to Islamic values and principles (Ahmad and Ahmad 2009; Fianto et al. 2018). Although conventional microfinance has similar objectives to Islamic financial principles, the operation of conventional MFIs conflict with Islamic financial principles (Alkhan and Hassan 2020) as microfinance generally deals with interest (riba) through microloans, conventional microfinance may in essence infringe on Islamic financial principles (Ahmed 2002).

Products offered by Islamic MFIs can be generally categorized into: (1) equity financing instruments, (2) credit or debt financing instruments and (3) other types of microfinancing such as investment deposits, mutual insurance schemes and savings accounts (Ahmad et al. 2020), Islamic MFIs also provide financial products that are aligned to Islamic principles such as profit and loss sharing (PLS) and non-PLS mechanisms (Dhumale and Sapcanin 1998). In general, Islamic MFIs generate profit through equity financing and debt-based financing (Fianto et al. 2018) whereby debt-based financing is the most common model of financing (Aggarwal and Yousef 2000; Asutay 2007; Dusuki and Abdullah 2006). In equity financing, Islamic MFIs share profits with their clients in a profit-sharing agreement whereby the parties involved share their resources in a project and share returns generated based on a pre-agreed ratio (Akhter et al. 2009; Abdul-Rahman et al. 2014). For debt-based financing (which is a non-PLS agreement), Islamic MFIs may earn a margin or fee from debt-based financing (Shahinpoor 2009).

Islamic microfinance products have also included products generally used in the Islamic Banking industry such as qardh hasan (benevolent loan), murabaha (cost-plus sale), ijara (lease), salam (forward sale), mudharaba (trust partnership), musharaka (joint venture partnership), among others (Hassan et al. 2013). Furthermore, the majority of Islamic MFIs offer two financial products: Murabaha and Qard-Hassan loans (El-Zoghbi and Tarazi 2013). Qard-Hassan loans are "benevolent loans" that are interest free (Fan et al. 2019) and under Murabaha financing, the financial institution purchases an asset on behalf of the entrepreneur, then the financial institution resells the asset to the entrepreneur at a predetermined price including the original cost and an additional profit margin, the entrepreneur repays the financial institution in lump sum or in future installments, then financial institution eventually transfers ownership to the entrepreneur when all payments are complete (Fan et al. 2019).

It has been found that religious identity plays an important factor for MFIs as religiously affiliated MFIs have lower financial sustainability compared to secular MFIs (Lensink et al. 2018). For instance, Mersland et al. (2013) argue that MFIs that are Christian based earn lower profits than conventional MFIs but they also incur lower funding costs. Studies conducted on Islamic MFIs found although they serve more borrowers and poorer borrowers therefore performing better in terms of increasing outreach, and in terms of financial results, some evidence was found indicating that Islamic MFIs underperform conventional MFIs (Ahmad et al. 2020), although the evidence is relatively weak.

Islamic Financial Inclusion

Many studies have found that faith and ethnicity impact financial literacy and financial decision-making (Guiso et al. 2002; Barro and Mitchell 2004; Gerrans et al. 2009; Khan and Khanna 2010; Demirgüç-Kunt and Klapper 2012a, b; Echchabi 2012; Naceur et al. 2017; Zulkhibri 2016; Hassan et al. 2018). Religiosity has been found to influence the acceptance of Islamic financial services (Amin and Aman 2016; Jaffar and Musa 2016; Maryam et al. 2019), particularly in Muslim-majority countries (Zulkhibri 2016; Hassan et al. 2018). Kim et al. (2020) examined the impact religiosity and social inequality had on financial inclusion and found that both in Muslim majority countries and non-Muslim majority countries, religious factors and social inequality (i.e., gender or educational attainment) had an impact on financial inclusion levels.

In order to enhance financial inclusion, financial products and services have to be accessible and affordable to all segments of society (Demirgüç-Kunt et al. 2018), product design and sophisticated documentation requirements in particular, are two significant barriers to financial inclusion (Beck et al. 2008). In Muslim-majority countries, the absence and uneven access to financial products and services that are aligned with Islamic principles could be a reason for the lack of bank accounts for adults in those countries (Zulkhibri 2016), therefore, to enhance financial inclusion, Islamic finance could be a way to bring those that are financially excluded, due to religious concerns, into the mainstream financial system (Jouti 2018).

According to Zulkhibri (2016), Islamic financial services could address the issue of financial inclusion in two ways, first by providing a suitable alternative to conventional debt-based financing through the promotion of risk-sharing contracts and second by using specific wealth distributive instruments, such as Islamic microfinance, to support the poor and enhance entrepreneurship. It is also important for Islamic financial institutions to offer products and services that are inexpensive, tailor-made to clients' needs, have greater collateral flexibility and have more straightforward documentation requirements (Ali et al. 2020).

Group-Based Lending

The success of providing finance to the poor is partly attributed to group-based lending mechanisms that allow the poor access to mainstream financial services that they traditionally did not have access to (Dusuki 2006). One of the main features of group-based lending is social capital which Putnam (1993) defines as social networks, norms and trust that facilitates mutual benefits through cooperation and coordination. According to Dusuki (2006), social capital is "the internal social and cultural coherence of society, imbuing qualities like sense of belonging, group loyalty, good will, sympathy, respect for others and teamwork among people and the institutions in which they are embedded."

The group lending approach allows those that are not included in the formal financial system to obtain financial services without physical collateral (Conning 1996; Hulme and Wright 2006). In other words, cooperation and trust allow those that are financially excluded to access capital by forming groups and associations that are based on shared goals. Group members are also expected to monitor and apply social pressure to ensure that every member that takes out a loan is obligated to repay the loan (Ritzer 2007) and peer pressure is used as a way to guarantee loan repayment as all members would be jointly liable for the loans that are provided to them (Dowla 2006). According to Nugroho and O'Hara (2008), two ways in which group members can be encouraged and be responsible to repay their loans are first by creating social punishment (such as rumors and bad reputation) for those that do not make any attempts to repay their loans and second through strengthening the moral value of the group through friendships. However, according to Eberhard (2008), the effectiveness of this approach depends on the trustworthiness of the members of the group.

Many scholars believe that in order to overcome the various barriers that have prevented large segments of the population from accessing formal financial institutions may entail more than financial intermediation in the conventional sense (Fukuyama 1995; Coleman 1988; Collier 2000). Integrating underserved groups into the mainstream financial system may require the building of social capital structures, such as upfront investments, to help develop the human resources of clients such as confidence, knowledge, skills and information (Dusuki 2006). The features of social capital can therefore be used by the poor in lieu of capital (Collier 2000) and as a way to create human capital (Coleman 1988). Besley and Coate (1995) argue that group-based lending can harness "social collateral" and could be a powerful incentive to yield higher repayment rates compared to individual lending. The social capital concept in group-based lending is also related to the Islamic notion of 'Asabiyah (social solidarity) that was mentioned by scholar Ibn Khaldun in his infamous book titled al-Muqaddimah (Dusuki 2006).

Savings groups are community-based financial institutions whereby individual savings are pooled together and kept in a safe place and then lent out to requesting members. They are considered as innovative mechanisms that bring financial inclusion to poor and vulnerable households who are not reached by mainstream banking or microfinance institutions and are fast spreading in sub-Saharan Africa and other developing countries (Burlando and Canidio 2017). They are a "type of community-owned microfinance intervention" with a focus on poor households (Tura et al. 2020). The savings group generates social capital in order to compensate for their lack of material assets and this in turn creates creditworthiness for borrowers and allows them to attract financing from financial institutions to fund their economic activities (Dusuki 2006). This is as the self-selected group members share the common interest of gaining access to credit and savings services and therefore have adequate low-cost information to screen other members of the group and apply sanctions to those who do not comply with the rules, and as a result this "lowers transaction costs, reduces financial risk and facilitates a greater range of market transactions in outputs, credit, land and labour which can in turn lead to better incomes" (Dusuki 2006).

CASE STUDY: PROGRAM FOR RESILIENT SYSTEMS (PROGRESS)

Case Context: Background on Wajir County

Located near the Somali border in northeastern Kenya, Wajir County is a sparsely populated region with a predominantly Muslim population with most residents clustered together living in small concentrated settlements or towns (Nübler et al. 2020). Wajir is one of the third largest counties in Kenya (out of 47) with a population of over 657,000 and covering an area of 55,840 km^2 (Sladkova 2019). In general, poverty rates are as high as 70% which is well above the national average of 45% in the northern Kenya region (World Bank 2018).

90% of residents in the region are heavily reliant on livestock raising as their main source of both food and income but in recent years, the pastoral lifestyle has been increasingly insecure given the severity of droughts and the reduced dependability on weather patterns (Tiwari et al. 2019). Drought and extreme weather conditions are not only a major factor of food scarcity in northern Kenya, but it also adversely impacts the livelihoods of rural communities; therefore, the inability to adapt to changing local climatic conditions coupled with low levels of resilience has increased hunger and malnutrition-related diseases, impacting highly vulnerable subsects of the population such as children and the elderly (Maione 2020).

Northern Kenyan regions, in particular, are highly prone to drought-risk resulting in heavy social and ecological burdens on people and ecosystems, such as overexploitation of fragile ecosystems and unsustainable agricultural practices (Jensen et al. 2014). Moreover, a compounding of these risks drives local pastoralist communities into deeper levels of poverty (Maione 2020). For instance, exposure to food insecurity and the pervasiveness of diseases that are malnutrition-related increase the likelihood of greater impoverishment (Barrett et al. 2008b).

Wajir county is a hot, semi-arid desert climate where the main economic activity of the county is pastoralism. Given that pastoralism, rather than crop agriculture, is the main source of income, livelihoods are very sensitive to low rainfall and drought conditions (Nübler et al. 2020) and these conditions can have significant effects on livestock prices and economic well-being (Kazianga and Udry 2006). Furthermore, low rainfall and drought lead to a decline in the availability of water which impacts livestock health and reduces prices (Opiyo et al. 2015) pastoralists are therefore particularly vulnerable as recovery from shocks—including having to rebuild herds—can take several years (Kimiti et al. 2018). In addition, in times of drought, pastoralist family members may need to travel far or even relocate to find suitable watering sites or pasture for their herds (MoALF 2017; Merttens 2018). Overall, droughts have become more frequent and intense in Wajir county, and it is predicted that drought intensity and frequency, as well as average annual temperatures, are expected to continue to rise, impacting agriculture and livestock production (Chaudhury et al. 2020). These anticipated changes threaten 85% of household income among nomadic pastoral communities (MoALF 2017).

Similar to many countries in sub-Saharan Africa, Kenya is characterized by a high dependency on the agricultural sector for revenues and employment and in general, smallholder farmers are the pillars of a lot of the economies of sub-Saharan African countries (Altieri 2009; Davis et al. 2017; GoK 2009; Salami et al. 2010). In Kenya, smallholder farms with areas ranging from 0.2 to 3 ha are the source of more than 70% of the country's total agriculture produce (Kamau et al. 2018), therefore, in a country where the agricultural sector is responsible for an estimated 26% of the gross domestic product (GDP), and 18% of formal and 60% of informal employment in rural areas (GoK 2009)—the role of smallholder farmers is essential. Hundreds of millions of smallholder farmers face serious climate-induced challenges such as poor and declining soil fertility, leading to large yield gaps for most crops (Kamau et al. 2018). Furthermore, as a result of pests and diseases, pre- and post-harvest crop and animal losses are high (GoK 2009; Salami et al. 2010; Tittonell and Giller 2013).

In general, smallholder farmers differ in structural aspects such as information access, asset availability and allocation as well as in functional aspects such as agricultural production objectives, livelihood strategies and their dynamics (Kuivanen et al. 2016; Pacini et al. 2014; Tittonell et al. 2010), diversification approaches (van de Steeg et al. 2010) and other social and economic aspects (Bidogeza et al. 2009). They also face financial challenges having limited access to financial capital and markets (Kamau et al. 2018). According to Kamau et al. (2018), given the heterogeneity of smallholder farmers in sub-Saharan Africa, any effort made to address their challenges needs to start with an understanding of their complex diversity. The mobility of pastoralists in Wajir and their access to resources was identified as the most important indication of resilience of local communities by humanitarian organization Mercy Corps as according to Mercy Corps, communities "are better able to cope with and absorb shocks and stresses through diverse livelihoods, responsive community institutions and community financing mechanisms" (Sladkova 2019).

Financial Inclusion in Kenya

In comparison with other countries in the region with similar income levels, Kenya's financial sector is larger, more diversified, further developed and has a higher level of financial inclusion (Weingärtner et al. 2019). Although financial inclusion in Kenya has grown tremendously over the years, it has nonetheless failed to provide adequate access to banking services for a large part of the population and in urban areas, lending is skewed in favor of large private and public enterprises (Ndegwa 2020). There are three main financial providers in Kenya, these are the savings and credit cooperatives, microfinance institutions and mobile money transfer services and as of 2013, only 27% of the Kenya population were financially included (Ali 2017). However, between 2013 and 2016, the number of people accessing a type of financial service through a formal or informal financial institution in Kenya increased from 74.9 to 82.6%—this was mainly driven largely by an expansion of mobile money (Weingärtner et al. 2019).

The Kenyan government recognized the importance of financial services in strengthening development and enhancing resilience; therefore, it engaged in a number of efforts to increase the adaptiveness of its large-scale social protection cash transfer programs (Weingärtner et al. 2019). An example as such is the Hunger Safety Net Program (HSNP) that operated in four drought-prone counties in northern Kenya, including Wajir. HSNP aimed to help vulnerable households cope with climate shocks (Maione 2020), however, social protection for the most food-insecure groups through targeted cash or input transfers can only address problems in the short-term. In the long-term, value-chain issues such as improved market linkages and more access to resources like land and capital need to be addressed (Kamau et al. 2018). In parallel to the HSNP, also operating in Wajir was the Program for Resilient Systems (PROGRESS) that was part of the Building Resilience and Adaptation to Climate Extremes and Disasters (BRACED) program led by Mercy Corps. The aim of PROGRESS was to enhance the resilience of pastoralist communities by providing them with access to Islamic financial services.

Building Resilience and Adaptation to Climate Extremes and Disasters (BRACED) Program

As regions in Kenya have extreme levels of vulnerability to the rising risk of climate-related shocks and stresses, a number of national strategies were created to address these risks, in addition to investments from bilateral and multilateral development agencies into large-scale programs like BRACED (Harvey et al. 2019). The BRACED program was originally a 3-year £110 million program funded by the United Kingdom's Department for International Development (Harvey et al. 2019) that was later extended to 2019. BRACED officially aimed at enhancing the resilience of up to 5 million vulnerable people against climate extremes and disasters in various parts of the world but with a focus on the Sahel furthermore, BRACED aimed at building the resilience of vulnerable rural populations as part of a wider, long-term development goal (Béné et al. 2020). For BRACED, resilience was understood as the so-called 3As approach—the capacity to (1) adapt to, (2) anticipate and (3) absorb, climate extremes and disasters (Bahadur et al. 2015). Moreover, the program aimed to enhance resilience to climate extremes and disasters as well as improve the integration of disaster risk reduction and climate adaptation efforts into development methods in Wajir and other places (BRACED 2016).

As BRACED was designed to facilitate large-scale investment into building the resilience of vulnerable populations, from 2015 to 2017, BRACED provided grants to 15 NGO-consortia that were working across 13 countries in East Africa, the Sahel and Asia (Harvey et al. 2019). The consortia invested in improving climate information acquisition and interpretation, increasing climate communication and information to end users and engaging the private sector as an alternative avenue to communicating climate information (BRACED 2017). Through the strengthening of climate services in the BRACED program, it was expected that community capacities were going to be strengthened when dealing with climate change shocks and stresses (Wilkinson et al. 2015).

Program for Resilient Systems (PROGRESS)

As part of the BRACED program, PROGRESS was implemented in Wajir with a special focus on three main areas; developing market systems; natural resource management and governance; and gender mainstreaming (Weingärtner et al. 2019). As for the market development component, a crucial element was to increase community engagement with the private sector to enhance access to financial and non-financial services and to enhance supplies and quality inputs with the outcome primarily focused on livestock markets, climate-smart agriculture/agro-ecology and clean energy products (Weingärtner et al. 2019). Overall the program aimed to build resilience against major climate-induced shocks and stresses such as droughts, deforestation, desertification, floods and soil erosion.

PROGRESS placed an emphasis on the delivery of financial services in northern Kenya as part of its efforts toward developing market systems in livestock markets, agriculture and clean energy, through cash savings and access to financing, as generally the project aimed to enhance resident's absorptive and adaptive capacities as well as contribute to growth (Weingärtner et al. 2019). As compared to other parts of the country, the penetration of formal and semi-formal financial services across northern and eastern Kenya is relatively low (FSD Kenya 2016); furthermore, in areas prone to seasonal drought events, the effects of droughts are widespread and have adverse ramifications on economic and agricultural systems and the natural environment, including loss of food availability and livestock assets (Wilkinson and Peters 2015). In this context, financial inclusion initiatives could potentially help enhance the absorptive capacity of households to prepare for and cope with shocks, take protective action and deal with extreme weather events.

PROGRESS provided pastoralist communities with access to Islamic compliant financial services through a savings and credit cooperative and support for Village Savings and Loans Associations (VSLAs) (Weingärtner et al. 2019). Although there were two fully fledged Islamic banks operating in Kenya at the time, the program aimed to address the lack of Islamic financial services by catering to the low-income share of the population who were unable to access banks due to higher fees and larger loan volumes (Weingärtner et al. 2019).

Prior to PROGRESS, the availability of adequate services in the predominantly Muslim county of Wajir was limited as no fully Islamic compliant formal financial institution had existed or operated there, therefore, for this reason, PROGRESS partnered with the first Shari'ah compliant savings and credit cooperative (SACCO) in Kenya known as Crescent Takaful SACCO (CTS) (Weingärtner et al. 2019). CTS provided a range of products including (Weingärtner et al. 2019):

- Ayuta Sokoni—which pools funds from entrepreneurs in order to support them in a variety of trade aspects such as restocking products,
- Ayuta Al-Rafiq—which enables groups of friends to come together to pool their resources and define their economic aims such as co-investments in business and,
- Mifugo Kash Kash—which aims to empower small-scale traders to boost their income by accessing markets outside Wajir and purchasing livestock.

The establishment of SACCOs in underserved areas and facilitating support services to enable connections between low-income groups and formal financial services providers has been a key role for PROGRESS (Weingärtner et al. 2019). In 2016, CTS opened two offices in Wajir, one in Wajir Town and the other in Habaswein, south of the county and by 2018, over 1,500 residents had started saving through CTS with just over 100 people taking out loans as individuals or as groups (Weingärtner et al. 2019). At that point, there were a total of 24 individual clients that had accessed 42 loans of an estimated KES 26,000 (on average close to GBP 2,000) and ten groups with an average number of seven members had taken out loans of almost KES 29,000 (close to just over GBP 2,200) (Weingärtner et al. 2019). Results from PROGRESS indicated that there is great potential for SACCOs such as CTS to cater to the needs and financial capabilities of residents in Wajir as they offer group lending approaches, lower administrative fees than banks and manageable loan amounts (Weingärtner et al. 2019). However, there were concerns about the financial viability of the strategy of CTS as in addition to a relatively high ratio of operational costs to loan value and fees, in order to service financially excluded groups, there needs to be adequate training on business skills and financial literacy to enhance capacity and trust—this would further increase costs to the financial institution (Weingärtner et al. 2019). There were also some trust issues and concerns about CTS. When surveyed, CTS clients said they trusted CTS because it was Shari'ah compliant but they complained about the hidden administrative costs involved in taking out loans and despite awareness, CTS continues to face challenges of mistrust and uncertainty regarding the compliance of its products and operations with Islamic finance principles (Weingärtner et al. 2019).

Lessons Learned

Further Product Development and Innovation

In terms of the project achieving its aim in developing market systems, to what extent the use of financial services, provided through CTS, helps to enhance climate-resilient economic development to Wajir residents cope with shocks is still yet to be seen (Weingärtner et al. 2019). It was found that a limited number of livestock herders or traders were accessing the standard CTS loans products that were available which according to Weingärtner et al. (2019), may "to some extent, be linked to the specific product design being less suited to the livestock business." Given that one of PROGRESS' main objectives was to enhance the residents of Wajir's absorptive and adaptive capacities, and contribute to growth, by developing market systems through the delivery of financial services, a lesson that can be learnt from the PROGRESS case is the need for further product development and innovation.

Moreover, although CTS appeared to have filled a gap by providing Islamic financial products in northern Kenya, the experience of PROGRESS revealed that there was a limited understanding of livestock trading and its impact on financial needs (Weingärtner et al. 2019). For instance, due to their nomadic lifestyle, a common pastoralist practice of saving is in the form of livestock as pastoralists anticipate more favorable returns when capital is held in livestock rather in savings (Ali 2017), which in turn could inhibit the uptake of financial services (Weingärtner et al. 2019). While focusing exclusively on providing funding to livestock traders has been a key entry point for the project (Weingärtner et al. 2019), PROGRESS' experience indicates that there is still a need for simpler financial products and services to be designed in thoughtful ways for the target population, and in particular for pastoralist communities.

Gender Mainstreaming Outcomes

According to Rao et al. (2019), in some semi-arid areas, vulnerability as a result of climatic and non-climatic factors are "inextricably linked, the severity of effect mediated by gender and wider social relations." Gender mainstreaming therefore considers the empowerment of women as a goal in itself and a crucial step in ensuring development interventions are effective and able to deliver their goals (Leder et al. 2017). Although development programming has tended to focus on the economic empowerment of individual women as an anti-poverty measure in order to enable them to contribute to national economic growth, according to Leder et al. (2017), it has failed to challenge the current model of development, in particular the technical and managerial model (Leder et al. 2017).

In many countries women are heavily involved in climate-sensitive activities such as farming, forestry and fisheries and have an abundance of experience and knowledge of environmental protection and adaptation strategies (Dankelman and Jansen 2010). Therefore, due to their substantial involvement in agriculture and their role in ensuring the well-being of their households, women have the potential to be key contributors in enhancing food security, livelihoods and resilience (Quisumbing et al. 1996). However, in spite of their contribution to general household livelihood strategies, in developing countries, women are often not included in efforts to improve resilience to climate shocks and disasters (McOmber et al. 2019). The lack of inclusion of women in activities to adapt to or mitigate climate risks in development efforts means there is untapped potential (Bob and Babugura 2014); therefore, beyond financing, effective approaches that address gender disparities and include gender-sensitive approaches could enhance resilience and adaptation to climate change.

Although the PROGRESS program operated under the premise that gender and other forms of social inclusion were key to an equitable climate change response, overall the final project evaluation struggled to demonstrate an explicit link between gender equality and resilience (Leavy et al. 2019). However, there were other important measures that indicated the project enabled the creation new spaces for women in Wajir to find mutual support, work together and build their confidence. For instance, PROGRESS organized women's networking events which facilitated peer-to-peer learning for women in the county to exchange knowledge and experiences about the challenges facing their businesses, enabling them to build coalitions to address challenges such as adverse climate change impacts and gendered practices (Leavy et al. 2019). While they may not directly contribute to the intended end outcomes of the project, the outcomes of these soft measures will most likely materialize in the long-term which in turn could help build community capacity and further disaster resilience. Going forward, such climate-resilient building projects that include gender equality outcomes should have a more detailed understanding of the specific gender dynamics of the region or the context in which the project is operating, in order to better draw a link between gender equality, resilience and adaptation to climate extremes and disasters.

FURTHER RECOMMENDATIONS

Building on the lessons learnt from PROGRESS, the following section outline more widely applicable lessons and further recommendations.

Increasing Trust

As mentioned above, it was found that there were some trust issues with CTS. Interviews conducted across three CTS focus groups found that although clients trusted CTS because it was Shari'ah compliant and did not charge them interest, there were some complaints about hidden administrative costs associated with taking out loans, which CTS had not clearly explained to clients when they were applying for loans (Weingärtner et al. 2019). Therefore, an ongoing challenge CTS continues to face is that of mistrust and uncertainty regarding the compliance of its products and operations with Islamic finance principles (Weingärtner et al. 2019) as there is a level of confusion of exactly how CTS products and services are Shari'ah complaint.

In order to increase trust and awareness of CTS, more capacity building and financial literacy initiatives could be implemented. In particular, initiatives should focus on increasing understanding of theory and current practices of Shari'ah compliant financial services in order to bridge the gap and improve access to finance. This is as the lack of knowledge of Shari'ah compliant financial services is not only leading to mistrust, but it also appears to be a significant barrier to financial inclusion and uptake of Shari'ah compliant products and services.

Also, it has been argued that combining microfinancial services with social, health and educational programs "are more effective in improving the welfare of the poor" (Abdul Rahman and Dean 2013) furthermore, Hassan et al. (2012) argue that such strategies can facilitate the achievement toward a "triple bottom line"—social, financial and spiritual well-being. Further research could be conducted to assess how such strategies can help expand access to financial services in northern Kenya in order to enhance resilience to climate change. As an integrated approach taking into consideration not just the financial but also social and spiritual elements could result in more uptake and trust in Shari'ah compliant financial services.

Digital Financial Tools

A potential solution for addressing many of the challenges of financial inclusion and closing the gap and "reaching last mile communities" is through digital technology (Tiwari et al. 2019). As digital tools can help overcome transparency issues that are associated with traditional financial services, they also have the potential to enhance autonomy over transactions and payments (Aker et al. 2016; Holloway et al. 2017).

For instance, mobile finance platform M-Pesa has been widely adopted throughout Kenya by at least one member of 96% of Kenyan households and has increased households' ability to save and protect themselves against shocks, provided users with an easy way to make and receive payments, and has given users access to broader social networks in times of crisis (Tiwari et al. 2019). This positive impact is due to mobile money accounts offering unbanked people the ability to send money back to their home villages faster, more cheaply and more securely increasing the stability of their revenues during tough times (Jouti 2018). Research has suggested that innovative tools such as M-Pesa have the potential to bridge the financial inclusion gap and spread the benefits of financial services as access to M-PESA lifted 194,000 Kenyan households out of poverty (Suri and Jack 2016). Going forward, further research on how the expansion and uptake of such digital financial tools can enhance resilience to climate change, should be conducted.

Linking Islamic Financial Products and Services

In recent years, index insurance has been introduced in a number of developing countries to help farmers and pastoralists deal with weather-related risks and poverty traps (Barnett et al. 2008; Jensen et al. 2018; Miranda and Farrin 2012; Smith 2016; Takahashi et al. 2016); however, demand for index insurance is generally low (Noritomo and Takahashi 2020). The literature on index insurance shows that it can benefit poor households through two channels (Bertram-Huemmer and Kraehnert 2018; Cole et al. 2017; Hill et al. 2019; Matsuda et al. 2019). The first is through the impact on resource allocation ex ante which then enables rural households to invest in higher risk and return activities and the second is the compensation of losses ex post through payouts, which can help rural households and communities recover from shocks at a quicker rate (Noritomo and Takahashi 2020).

Index-based livestock insurance aims to protect pastoralists who herd camels, cattle, goats and sheep, from drought-related forage scarcity (Johnson et al. 2019). Northern Kenya is at the forefront of index-based livestock Takaful (Shari'ah compliant insurance) development, with both commercial products sold through Takaful Africa and subsidized insurance under the Kenyan Livestock Insurance Program (KLIP) being available to livestock owners in some part of Wajir (Weingärtner et al. 2019). In comparison with conventional index-based livestock insurance, the index on which the contract for a specialized Kenyan Takaful company is settled does not differ, however, what separates them is the internal financial structure whereby a Mudaraba model is used—separating the risk fund that was being used to pay claims from the fund being used to pay shareholder dividends (Johnson et al. 2019). In order to further develop market systems, Shari'ah compliant financial products and services that address and tailor to the unique needs of pastoralists and the livestock business could go a long way. Therefore, further research should be conducted on the potential and viability of linking CTS products and services with index-based livestock Takaful products in order to enhance resilience to climate change.

Conclusion

Increasing access to financial services is part of five of the 17 Sustainable Development Goals (SDGs) set by the United Nations for 2030 and has been a top priority for policymakers around the world. As extreme weather-related shocks such as droughts, floods, deforestation, desertification, soil erosion and water and food insecurity are more frequent in the developing world—the role of financial services in enhancing climate resilience has been of interest in recent years.

In the northeast of Kenya, recurrent droughts and other extreme climate change induced events undermine the ability of the populations living in those regions to sustain shocks and stresses which results in increased vulnerability to humanitarian crises and poverty traps. The aim of this chapter, therefore, was to evaluate PROGRESS that was part of the wider BRACED program implemented by humanitarian organization Mercy Corps in the Kenyan county of Wajir, to assess whether expanding access to Islamic financial services can increase resilience and adaptation to climate change.

It was found that there was great potential for SACCOs to cater to the financial needs of residents in Wajir as they offered Islamic financial products and services as well as group lending options. As there is a large Muslim population in Wajir, there was a high demand for tailored Islamic financial services in order to comply to spiritual needs. However, the extent to which the use of Islamic compliant products and services provided through the program helped to enhance the climate resilience of Wajir residents is inconclusive. The program faced challenges of mistrust and uncertainty regarding the Shari'ah compliance of the financial services provided also, in regard to gender mainstreaming outcomes, the program did not demonstrate a clear link between greater gender equality and resilience.

According to Béné et al. (2020), one of the key principles that underpin recent conceptualizations of resilience is "the recognition that resilience should not be seen as the final goal of a development program but instead as an intermediate outcome required for achievement of a more fundamental goal related to a longer-term developmental ambition, typically a measure of wellbeing (e.g. food security, health/nutrition status, poverty reduction)." In other words, resilience should not be considered as the ultimate objective of a program and rather an intermediary byproduct. The goal of humanitarian and development programs and projects implemented should be to improve the long-term well-being and welfare of people (Béné et al. 2015; Constas et al. 2014). Given resilience programming—as a method of addressing and tackling the vulnerability of populations living areas that are prone to frequent climate change shock and stresses—is gaining more interest, going forward, further research should be conducted on the potential of digital financial solutions and the potential of linking Islamic financial products and services to enhance resilience to climate change.

References

Abdul Rahman, A. R. (2007). Islamic microfinance: A missing component in Islamic banking. *Kyoto Bulletin of Islamic Area Studies*, 1(2), 38–53.

Abdul-Rahman, A., Latif, R. A., Muda, R., and Abdullah, M. A. (2014). Failure and potential of profit-loss sharing contracts: A perspective of New Institutional, Economic (NIE) Theory. *Pacific-Basin Finance Journal*, 28, 136–151.

Adegbite, O. O., and Machethe, C. L. (2020). Bridging the financial inclusion gender gap in smallholder agriculture in Nigeria: An untapped potential for sustainable development. *World Development*, 127, 104755.

Adeola, O., and Evans, O. (2017). The impact of microfinance on financial inclusion in Nigeria. The Journal of Developing Areas, 51(4), 193–206.

Aggarwal, R. K., and Yousef, T. (2000). Islamic banks and investment financing. *Journal of Money, Credit and Banking*, 93–120.

Aghion, P., and Bolton, P. (1997). A theory of trickle-down growth and development. *The Review of Economic Studies*, 64(2), 151–172.

Agyemang-Badu, A. A., Agyei, K., and Kwaku Duah, E. (2018). Financial inclusion, poverty and income inequality: Evidence from Africa. *Spiritan International Journal of Poverty Studies*, 2(2).

Ahmad, A. U. F., and Ahmad, A. R. (2009). Islamic microfinance: The evidence from Australia. *Humanomics*, 25(3), 217–235.

Ahmad, S., Lensink, R., and Mueller, A. (2020). The double bottom line of microfinance: A global comparison between conventional and Islamic microfinance. *World Development*, 136, 105130.

Ahmed, H. (2002). Financing microenterprises: An analytical study of Islamic microfinance institutions. *Islamic Economic Studies*, 9(2), 27–64.

Ajefu, J. B., Demir, A., and Haghpanahan, H. (2020). The impact of financial inclusion on mental health. *SSM-Population Health*, 11, 100630.

Aker, J. C., Boumnijel, R., McClelland, A., and Tierney, N. (2016). Payment mechanisms and antipoverty programs: Evidence from a mobile money cash transfer experiment in Niger. *Economic Development and Cultural Change*, 65(1), 1–37.

Akhter, W., Akhtar, N., and Jaffri, S. K. A. (2009). Islamic micro-finance and poverty alleviation: A case of Pakistan. Proceeding of the 2nd CBRC, Lahore, 1–8.

Alano, E., and Lee, M. (2016). Natural disaster shocks and macroeconomic growth in Asia: Evidence for typhoons and droughts. Economics Working Paper Series, 503. Manila: Asian Development Bank.

Alkhan, A. M. (2016). A critical analysis of special purpose vehicles in the Islamic banking industry: The Kingdom of Bahrain as a case study (Doctoral dissertation, University of Bolton).

Ali, A. E. E. S. (2017). The challenges facing poverty alleviation and financial inclusion in North-East Kenya Province (NEKP). *International Journal of Social Economics*, 44(12), 2208–2223.

Ali, M. M., Devi, A., Furqani, H., and Hamzah, H. (2020). Islamic financial inclusion determinants in Indonesia: an ANP approach. *International Journal of Islamic and Middle Eastern Finance and Management*, 13(4), 727–747.

Alkhan, A. M., and Hassan, M. K. (2020). Does Islamic microfinance serve maqāsid al-shari'a? Borsa Istanbul Review.

Allen, F., Demirguc-Kunt, A., Klapper, L., and Peria, M. S. M. (2012). The foundations of financial inclusion: Understanding ownership and use of formal accounts. *Journal of Financial Intermediation*, 27, 1–30.

Altieri, M. A. (2009). Agroecology, small farms, and food sovereignty. *Monthly Review*, 61(3), 102–113.

Ambarkhane, D., Singh, A. S., and Venkataramani, B. (2016). Measuring financial inclusion of Indian states. *International Journal of Rural Management*, 12(1), 72–100.

Amin, K., and Aman, Q. (2016). Determinants of attitude towards the acceptance of Islamic banking: A case of district Peshawar, Pakistan. *Journal of Managerial Sciences*, 10(1), 140–155.

Armendáriz, B., and Morduch, J. (2010). *The economics of microfinance*. MIT Press.

Asutay, M. (2007). Conceptualisation of the second best solution in overcoming the social failure of Islamic finance: Examining the overpowering of homoislamicus by homoeconomicus. *IIUM Journal in Economics and Management*, 15(2), 167–195.

Bahadur, A. V., Peters, K., Wilkinson, E., Pichon, F., Gray, K., and Tanner, T. (2015). The 3As: Tracking resilience across BRACED. Working and discussion papers. London: Overseas Development Institute (ODI).

Banerjee, A., and A. Newman. (1993). Occupational choice and the process of development. *Journal of Political Economy*, 101(2), 274–298.

Baquero, G., Hamadi, M., and Heinen, A. (2018). Competition, loan rates, and information dispersion in nonprofit and for-profit microcredit markets. *Journal of Money, Credit and Banking*, 50(5), 893–937.

Barnett, B. J., Barrett, C. B., and Skees, J. R. (2008). Poverty traps and index-based risk transfer products. *World Development*, 36(10), 1766–1785.

Barrett, C. B., Carter, M. R., Chantarat, S., McPeak, J. G., and Mude, A. G. (2008). Altering poverty dynamics with index insurance: Northern Kenya's HSNP.

Barro, R. and Mitchell, J. (2004). *Religious faith and economic growth: What matters most – belief or belonging?* Washington, DC: Heritage Foundation.

Beck, T., Levine, R., and Loayza, N. (2000). Finance and the sources of growth. *Journal of Financial Economics*, 58(1–2), 261–300.

Beck, T., Demirgüç-Kunt, A., and Levine, R. (2007a). Finance, inequality and the poor. *Journal of Economic Growth*, 12(1), 27–49.

Beck, T., A. Demirguc-Kunt, and M. S. M. Peria. (2007b). Reaching out: Access to and use of banking services across countries. *Journal of Financial Economics*, 85(1), 234–66.

Beck, T., Demirgüç-Kunt, A., and Martinez Peria, M. S. (2008). Banking services for everyone? Barriers to bank access and use around the world. *The World Bank Economic Review*, 22(3), 397–430.

Beck, T., Demirgüç-Kunt, A., and Honohan, P. (2009). Access to financial services: Measurement, impact, and policies. *The World Bank Research Observer*, 24(1), 119–145.

Béné, C., Frankenberger, T., and Nelson, S. (2015). Design, monitoring and evaluation of resilience interventions: conceptual and empirical considerations. Brighton: Institute of Development Studies. Retrieved from: https://opendocs.ids.ac.uk/opendocs/handle/20.500.12413/6556.

Béné, C., Riba, A., and Wilson, D. (2020). Impacts of resilience interventions–evidence from a quasi-experimental assessment in Niger. *International Journal of Disaster Risk Reduction*, 43, 101390.

Berger, A. N., Miller, N. H., Petersen, M. A., Rajan, R. G., and Stein, J. C. (2005). Does function follow organizational form? Evidence from the lending practices of large and small banks. *Journal of Financial Economics*, 76(2), 237–269.

sBerger, A. N., Frame, W. S. and Ioannidou, V. (2011). Tests of ex ante versus ex post theories of collateral using private and public information. *Journal of Financial Economics*, 100(1), 85–97.

Berhane, G., and Gardebroek, C. (2011). Does microfinance reduce rural poverty? Evidence based on household panel data from northern Ethiopia. *American Journal of Agricultural Economics*, 93(1), 43–55.

Bertram-Huemmer, V., and Kraehnert, K. (2018). Does index insurance help households recover from disaster? Evidence from IBLI Mongolia. *American Journal of Agricultural Economics*, 100(1), 145–171.

Besley, T. and Coate, S. (1995). Group lending, repayment incentives and social collateral. *Journal of Development Economics*, 46, 1–18.

Bidogeza, J. C., Berentsen, P. B. M., De Graaff, J., and Lansink, A. O. (2009). A typology of farm households for the Umutara Province in Rwanda. *Food Security*, 1(3), 321–335.

Boatright, J. R. (2013). *Ethics in finance*. Wiley.

Borgerhoff Mulder, M., Fazzio, I., Irons, W., McElreath, R. L., Bowles, S., Bell, A., Hertz, T., and Hazzah, L. (2010). Pastoralism and wealth inequality: Revisiting an old question. *Current Anthropology*, 51(1), 35–48.

Botzen, W. W., Deschenes, O., and Sanders, M. (2019). The economic impacts of natural disasters: A review of models and empirical studies. *Review of Environmental Economics and Policy*, 13(2), 167–188.

Bob, U., and Babugura, A. (2014). Contextualising and conceptualising gender and climate change in Africa. *Agenda*, 28(3), 3–15.

Boyd, E., Cornforth, R. J., Lamb, P. J., Tarhule, A., Lélé, M. I., and Brouder, A. (2013). Building resilience to face recurring environmental crisis in African Sahel. *Nature Climate Change*, 3(7), 631–637.

BRACED. (2017). BRACED resilience exchange: What have we learned so far? www.braced-rx.org.

Burlando, A., and Canidio, A. (2017). Does group inclusion hurt financial inclusion? Evidence from ultra-poor members of Ugandan savings groups. *Journal of Development Economics*, 128, 24–48.

Carbo S., Gardener E. P., and Molyneux, P. (2005). *Financial exclusion*. Palgrave MacMillan.

Carr, E. R. (2019). Properties and projects: Reconciling resilience and transformation for adaptation and development. *World Development*, 122, 70–84.

Cavallo, E., Galiani, S., Noy, I., and Pantano, J. (2013). Catastrophic natural disasters and economic growth. *Review of Economics and Statistics*, 95(5), 1549–1561.

Chaudhury, M., Summerlin, T., and Ginoya, N. (2020). Mainstreaming climate change adaptation in Kenya: Lessons from Makueni and Wajir counties. Working Paper. World Resources Institute. Retrieved from: http://www.indiaenvironmentportal.org.in/files/file/mainstreaming-climate-change-adaptation-kenya.pdf.

Chikalipah, S. (2018). Credit risk in microfinance industry: Evidence from sub-Saharan Africa. *Review of Development Finance*, 8(1), 38–48.

Chong, B. S., and Liu, M. H. (2009). Islamic banking: interest-free or interest-based? *Pacific-Basin Finance Journal*, 17(1), 125–144.

Cihak, M., Mare, D. S., and Melecky, M. (2016). The nexus of financial inclusion and financial stability: A study of trade-offs and synergies. Policy Research Working Paper, no. 7722. Washington, DC: World Bank Group.

Cole, S., Giné, X., and Vickery, J. (2017). How does risk management influence production decisions? Evidence from a field experiment. *The Review of Financial Studies*, 30(6), 1935–1970.

Coleman, J. S. (1988). Social capital in the creation of human capital. *American Journal of Sociology*, 94, 95–120.

Collier, P. (2000). *Social capital and poverty*. Washington, DC: The World Bank.

Conning, J. (1996). Group lending, moral hazard, and the creation of social collateral. Center for Institutional Reform and the Informal Sector, University of Maryland at College Park.

Constas, M., Frankenberger, T., Hoddinott, J., Mock, N., Romano, D., Bene, C., and Maxwell, D. (2014). A common analytical model for resilience measurement: causal framework and methodological options. Resilience Measurement Technical Working Group, FSiN Technical Series Paper, 2, 52.

Cull, R., Demirgüç-Kunt, A., and Morduch, J. (2009). Microfinance meets the market. *Journal of Economic Perspectives*, 23(1), 167–192.

Cumming, D., Johan, S., and Zhang, M. (2014). The economic impact of entrepreneurship: Comparing international datasets. *Corporate Governance: An International Review*, 22(2), 162–178.

Dahal, M., and Fiala, N. (2020). What do we know about the impact of microfinance? The problems of statistical power and precision. *World Development*, 128, 104773.

Dankelman, I., and W. Jansen (2010). Gender, environment, and climate change: Understanding the linkages. In: *Gender and climate change: An introduction* (pp. pp. 21–54). London : Earthscan.

Davis, B., Di Giuseppe, S., and Zezza, A. (2017). Are African households (not) leaving agriculture? Patterns of households' income sources in rural Sub-Saharan Africa. *Food Policy*, 67, 153–174.

de Oliveira Leite, R., dos Santos Mendes, L., and de Lacerda Moreira, R. (2020). Profit status of microfinance institutions and incentives for earnings management. *Research in International Business and Finance*, 54, 101255.

Demirgüç-Kunt, A., and Levine, R. (2009). Finance and inequality: Theory and evidence. *Annual Review of Financial Economics*, 1(1), 287–318.

Demirguc-Kunt, A., and Klapper, L. (2012a). Measuring financial inclusion: The global findex database. World Bank Policy Research Paper 6025. Washington, DC: The World Bank.

Demirgüç-Kunt, A., and Klapper, L. (2012b). Financial inclusion in Africa. Policy Research Working Paper. The World Bank Development Research Group Finance and Private Sector Development Team, 1–18.

Demirgüç-Kunt, A., Klapper, L. F., Singer, D., and Van Oudheusden, P. (2015). The global findex database 2014: Measuring financial inclusion around the world. World Bank Policy Research Working Paper, 7255.

Demirguc-Kunt, A., L. Klapper, D. Singer, S. Ansar, and J. Hess (2018). The global findex database 2017: Measuring financial inclusion and the fintech revolution. The World Bank.

Dhumale, R., and Sapcanin, A. (1998). An application of islamic banking principles to microfinance. Technical Note, A Study by the Regional Bureau for Arab States, UNDP, in Cooperation with the Middle East and North Africa Region, World Bank.

Dowla, A. (2006). In credit we trust: Building social capital by Grameen Bank in Bangladesh. *The Journal of Socio-Economics*, 35(1), 102–122.

Dusuki, A. W. (2006). Empowering Islamic microfinance: lesson from group-based lending scheme and Ibn Khaldun's concept of 'Asabiyah. In 4th International Islamic Banking and Finance Conference, Monash University Kuala Lumpur.

Dusuki, A. W., and Abdullah, I. N. (2006). The ideal of Islamic banking: chasing a mirage. In: *INCEIF Islamic banking and finance educational colloquium, Bank Negara Malaysia, Kuala Lumpur* (Vol. 3).

Duvendack, M., Palmer-Jones, R., Copestake, J. G., Hooper, L., Loke, Y., and Rao, N. (2011). What is the evidence of the impact of microfinance on the well-being of poor people? EPPI-Centre, Social Science Research Unit, Institute of Education, University of London, London.

Eberhard, J. W. (2008). Attaining empowerment: The potential of religious social capital. Working paper. Georgia State University, Atlanta, GA.

Echchabi, A. (2012). The relationship between religiosity and customers' adoption of Islamic banking services in Morocco. *Oman Chapter of Arabian Journal of Business and Management Review*, 1(10), 1–6.

El-Zoghbi, M., and Tarazi, M. (2013). Trends in Sharia-compliant financial inclusion. Focus Note, 84, 1–11. Washington, DC: Consultative Group to Assist the Poor (CGAP).

Fan, Y., John, K., Liu, F. H., and Tamanni, L. (2019). Security design, incentives, and Islamic microfinance: Cross country evidence. *Journal of International Financial Markets, Institutions and Money*, 62, 264–280.

Fianto, B. A., Gan, C., Hu, B., and Roudaki, J. (2018). Equity financing and debt-based financing: Evidence from Islamic microfinance institutions in Indonesia. *Pacific-Basin Finance Journal*, 52, 163–172.

Felbermayr, G., and Gröschl, J. (2014). Naturally negative: The growth effects of natural disasters. *Journal of Development Economics*, 111, 92–106.

Ferreira, S., and Karali, B. (2015). Do earthquakes shake stock markets? *PloS One*, 10(7), e0133319.

Fraser, N. (2009). Feminism, Capitalism, and the Cunning of History. *New Left Review*, 56, 97–117.

Fukuyama, F. (1995). Social capital and the global economy. *Foreign Affairs*, 74(5), 89–98.

FSD Kenya (2016). Data on financial service provision. Nairobi: FSD Kenya. Retrieved from: https://fsdkenya.org/knowledge-hub/dataset/.

Galor, O., and J. Zeira. (1993). Income distribution and macroeconomics. *The Review of Economic Studies*, 60(1), 35–52.

Garmaise, M. J. and Natividad, G. (2010). Information, the Cost of Credit, and Operational Efficiency: An Empirical Study of Microfinance. *Review of Financial Studies*, 23(6), 2560–2589.

Garmaise, M. J. and Natividad, G. (2013). Cheap credit, lending operations, and international politics: The case of global microfinance. *Journal of Finance*, 68(4), 1551–1576.

Gerrans, P., Clark-Murphy, M., and Truscott, K. (2009). Financial literacy and superannuation awareness of indigenous Australians: pilot study results. *Australian Journal of Social Issues*, 44(4), 417–439.

Goldman, M. J., and Riosmena, F. (2013). Adaptive capacity in Tanzanian Maasailand: Changing strategies to cope with drought in fragmented landscapes. *Global Environmental Change*, 23(3), 588–597.

Government of Kenya (GoK). (2009). Agricultural sector development strategy (ASDS), 2009–2020.

Government of Kenya, (GOK). (2015). Unlocking our full potential for realization of the Kenya Vision 2030. National Policy for the Sustainable Development of Northern Kenya and Other Arid Lands. Ministry of Devolution and Planning.

Grist, N., Mosello, B., Roberts, R., and Hilker, L. M. (2014). Resilience in the Sahel: Building better lives with children. Plan, Surrey.

Guérin, I., Labie, M., and Servet, J. (2015). *The crises of microcredit*. London: Zed Books.

Guiso, L., Haliassos, M., and Jappelli, T. (Eds). (2002). Household Portfolios, MIT Press.

Haan, N., Devereux, S., and Maxwell, D. (2012). Global implications of Somalia 2011 for famine prevention, mitigation and response. *Global Food Security*, 1(1), 74–79.

Hariharan, G., and Marktanner, M. (2012). The growth potential from financial inclusion. *ICA Institute and Kennesaw State University*, 2(5), 1–12.

Harvey, B., Jones, L., Cochrane, L., and Singh, R. (2019). The evolving landscape of climate services in sub-Saharan Africa: What roles have NGOs played? *Climatic Change*, 157(1), 81–98.

Hassan, S., Abdul Rahman, R., Abu Bakar, N., and Lahsasna, A. (2012). Towards triple bottom lines microfinance institutions (MFIs) – A case study of Amanah Ikhtiar Malaysia (AIM). In: Proceedings from International Conference on Excellence in Business, University of Sharjah, United Arab Emirates.

Hassan, M. K., Kayed, R. N., and Oseni, U. A. (2013). *Introduction to Islamic banking & finance*. Pearson Education Limited.

Hassan, M. K., Hossein, S., and Unsal, O. (2018). Religious preference and financial inclusion: The case for Islamic finance. In: *Management of Islamic finance: Principle, practice, and performance*. Emerald Publishing Limited.

Hermes, N., Lensink, R., and Meesters, A. (2011). Outreach and efficiency of microfinance institutions. *World Development*, 39(6), 938–948.

Hill, R. V., Kumar, N., Magnan, N., Makhija, S., de Nicola, F., Spielman, D. J., and Ward, P. S. (2019). Ex ante and ex post effects of hybrid index insurance in Bangladesh. *Journal of Development Economics*, 136, 1–17.

Hobbs, C., Gordon, M., and Bogart, B. (2012). When business is not as usual: decision-making and the humanitarian response to the famine in South Central Somalia. *Global Food Security*, 1(1), 50–56.

Holloway, K., Z. Niazi, and Rouse, R. (2017). *Women's economic empowerment through financial inclusion: A review of existing evidence and remaining knowledge gaps*. New Haven: Innovations for Poverty Action.

Honohan, P. (2004). Financial development, growth and poverty: How close are the links? In: *Financial development and economic growth* (pp. 1–37). London: Palgrave Macmillan.

Honohan, P. (2008). Cross-country variation in household access to financial services. *Journal of Banking & Finance*, 32(11), 2493–2500.

Huang, Y., and Zhang, Y. (2020). Financial inclusion and urban–rural income inequality: Long-run and short-run relationships. *Emerging Markets Finance and Trade*, 56(2), 457–471.

Hulme, M. K. and Wright C. (2006). Internet based social lending: Past, present and future. *Social Futures Observatory*, 11, 1–115.

IPCC. (2018). Global warming of 1.5 °C. An IPCC special report on the impacts of global warming of 1.5 °C above pre-industrial levels and related global greenhouse gas emission pathways, in the context of strengthening the global response to the threat of climate change, sustainable development, and efforts to eradicate poverty. Geneva: Intergovernmental Panel on Climate Change (IPCC).

Jaffar, M. A., and Musa, R. (2016). Determinants of attitude and intention towards Islamic financing adoption among non-users. *Procedia Economics and Finance*, 37(16), 227–233.

Jensen, N., Barrett, C. B., and Mude, A. (2014). Index insurance and cash transfers: A comparative analysis from northern Kenya.

Jensen, N. D., Mude, A. G., and Barrett, C. B. (2018). How basis risk and spatiotemporal adverse selection influence demand for index insurance: Evidence from northern Kenya. *Food Policy*, 74, 172–198.

Johnson, L., Wandera, B., Jensen, N., and Banerjee, R. (2019). Competing expectations in an index-based livestock insurance project. *The Journal of Development Studies*, 55(6), 1221–1239.

Jouti, A. T. (2018). Islamic finance: Financial inclusion or migration? *ISRA International Journal of Islamic Finance*, 10(2), 277–288.

Kamau, J. W., Stellmacher, T., Biber-Freudenberger, L., and Borgemeister, C. (2018). Organic and conventional agriculture in Kenya: A typology of smallholder farms in Kajiado and Murang'a counties. *Journal of Rural Studies*, 57, 171–185.

Karim, N., Tarazi, M., and Reille, X. (2008). Islamic microfinance: An emerging market Niche. Washington, DC: Consultative Group to Assist the Poor (CGAP).

Karim, L. (2011). *Microfinance and its discontents: Women in debt in Bangladesh.* Minneapolis, MN: University of Minnesota Press.

Karim, N., and Khaled, M. (2011). *Taking Islamic microfinance to scale.* Washington, DC: Consultative Group to Assist the Poor (CGAP).

Kazianga, H., and Udry, C. (2006). Consumption smoothing? Livestock, insurance and drought in rural Burkina Faso. *Journal of Development Economics*, 79(2), 413–446.

Khan, M. S. (1986). Islamic interest-free banking: a theoretical analysis. *International Monetary Fund Staff Papers*, 33(1), 1–27.

Khan, A. K., and Khanna, T. (2010). *Is faith a luxury for the rich? Examining the influence of religious beliefs on individual financial choices.* Harvard Business School.

Kim, J. H. (2016). A study on the effect of financial inclusion on the relationship between income inequality and economic growth. *Emerging Markets Finance and Trade*, 52(2), 498–512.

Kim, D. W., Yu, J. S., and Hassan, M. K. (2020). The influence of religion and social inequality on financial inclusion. *The Singapore Economic Review*, 65(1), 193–216.

Kimiti, K. S., Western, D., Mbau, J. S., and Wasonga, O. V. (2018). Impacts of long-term land-use changes on herd size and mobility among pastoral households in Amboseli ecosystem, Kenya. *Ecological Processes*, 7(1), 1–9.

Klapper, L., Laeven, L., and Rajan, R. (2006). Entry regulation as a barrier to entrepreneurship. *Journal of Financial Economics*, 82(3), 591–629.

Kling, G., Volz, U., Murinde, V., and Ayas, S. (2021). The impact of climate vulnerability on firms' cost of capital and access to finance. *World Development*, 137, 105131.

Kuivanen, K. S., Alvarez, S., Michalscheck, M., Adjei-Nsiah, S., Descheemaeker, K., Mellon-Bedi, S., and Groot, J. C. (2016). Characterising the diversity of smallholder farming systems and their constraints and opportunities for innovation: A case study from the Northern Region, Ghana. *NJAS-Wageningen Journal of Life Sciences*, 78, 153–166.

Lautze, S., Bell, W., Alinovi, L., and Russo, L. (2012). Early warning, late response (again): The 2011 famine in Somalia. *Global Food Security*, 1(1), 43–49.

Leavy, J., Boydell, E., McDowell, S., and Sladkova, B. (2019). Resilience results: BRACED final evaluation. BRACED Knowledge Manager. Retrieved from: https://reliefweb.int/sites/reliefweb.int/files/resources/resilience_results-_braced_final_evaluation.pdf.

Leder, S., Clement, F., and Karki, E. (2017). Reframing women's empowerment in water security programmes in Western Nepal. *Gender & Development*, 25(2), 235–251.

Ledgerwood, J., Earne, J., and Nelson, C. (Eds.). (2013). *The new microfinance handbook: A financial market system perspective*. World Bank Publications.

Lensink, R., Mersland, R., Vu, N. T. H., and Zamore, S. (2018). Do microfinance institutions benefit from integrating financial and nonfinancial services?. *Applied Economics*, 50(21), 2386–2401.

Leyshon, A., and Thrift, N. (1995). Geographies of financial exclusion: Financial abandonment in Britain and the United States. *Transactions of the Institute of British Geographers*, 312–341.

Liñares-Zegarra, J., and Wilson, J. O. (2018). The size and growth of microfinance institutions. *The British Accounting Review*, 50(2), 199–213.

Littlefield, E., Morduch, J., and Hashemi, S. (2003). Is microfinance an effective strategy to reach the millennium development goals? *Focus Note*, 24, 1–11.

Mader, P. (2015). The financialization of poverty. In: *The political economy of microfinance*. London: Palgrave Macmillan.

Mader, P. (2018). Contesting financial inclusion. *Development and Change*, 49(2), 461–483.

Maione, C. (2020). Adapting to drought and extreme climate: Hunger Safety Net Programme, Kenya. *World Development Perspectives*, 20, 100270.

Majid, N., and McDowell, S. (2012). Hidden dimensions of the Somalia famine. *Global Food Security*, 1(1), 36–42.

Maryam, S. Z., Mehmood, M. S., and Khaliq, C. A. (2019). Factors influencing the community behavioral intention for adoption of Islamic banking. *International Journal of Islamic and Middle Eastern Finance and Management*, 12(4).

Matsuda, A., Takahashi, K., and Ikegami, M. (2019). Direct and indirect impact of index-based livestock insurance in Southern Ethiopia. *The Geneva Papers on Risk and Insurance-Issues and Practice*, 44(3), 481–502.

McOmber, C., Audia, C., and Crowley, F. (2019). Building resilience by challenging social norms: Integrating a transformative approach within the BRACED consortia. *Disasters*, 43, S271–S294.

Mendelsohn, R., Emanuel, K., Chonabayashi, S., and Bakkensen, L. (2012). The impact of climate change on global tropical cyclone damage. *Nature Climate Change*, 2(3), 205–209.

Mersland, R., D'espallier, B., and Supphellen, M. (2013). The effects of religion on development efforts: Evidence from the microfinance industry and a research agenda. *World Development*, 41, 145–156.

Merttens, F. (2018).*Evaluation of the Kenya hunger safety net programme phase 2: Impact evaluation final report.* Oxford, UK: Oxford Policy Management.

Ministry of Agriculture, Livestock and Fisheries of Kenya (MoALF). (2017). Climate risk profile for Wajir County. Kenya County Climate Risk Profile Series. Nairobi, Kenya: The Ministry of Agriculture, Livestock and Fisheries (MoALF).

Miranda, M. J., and Farrin, K. (2012). Index insurance for developing countries. *Applied Economic Perspectives and Policy*, 34(3), 391–427.

Mohan, R. (2006). Economic growth, financial deepening and financial inclusion. Reserve Bank of India Bulletin, 1305.

Morduch, J. (1999). The microfinance promise. *Journal of Economic Literature*, 37(4), 1569–1614.

Morduch, J. (2000). The microfinance schism. *World Development*, 28(4), 617–629.

Morgan, P., and Pontines, V. (2014). *Financial stability and financial inclusion.* Asian Development Bank Institute.

Naceur, S. B., Barajas, A., and Massara, A. (2017). *Can Islamic banking increase financial inclusion? Handbook of empirical research on Islam and economic life.* Edward Elgar Publishing.

Ndegwa, R. (2020). The role of communication in SACCOs In promoting financial inclusion. *Kenya. European Journal of Economic and Financial Research*, 4(2).

Neaime, S., and Gaysset, I. (2018). Financial inclusion and stability in MENA: Evidence from poverty and inequality. *Finance Research Letters*, 24, 230–237.

Noritomo, Y., and Takahashi, K. (2020). Can insurance payouts prevent a poverty trap? Evidence from randomised experiments in northern Kenya. *The Journal of Development Studies*, 56(11), 2079–2096.

Nübler, L., Austrian, K., Maluccio, J. A., and Pinchoff, J. (2020). Rainfall shocks, cognitive development and educational attainment among adolescents in a drought-prone region in Kenya. *Environment and Development Economics*, 1–22.

Nugroho, A. E., and O'Hara, P. A. (2008). Microfinance sustainability and poverty outreach: a case study of microfinance and social capital in rural Java, Indonesia. Working paper. University of Technology GPO, Perth, Australia.

Obaidullah, M. (2008). Introduction to Islamic Microfinance. India: International Institute of Islamic Business and Finance.

Opiyo, F., Wasonga, O., Nyangito, M., Schilling, J., and Munang, R. (2015). Drought adaptation and coping strategies among the Turkana pastoralists of northern Kenya. *International Journal of Disaster Risk Science*, 6(3), 295–309.

Opondo, M., Abdi, U., and Nangiro, P. (2016). Assessing gender in resilience programming: Uganda. BRACED Resilience Intel, 2(2). Retrieved from: https://assets.publishing.service.gov.uk/media/57a08964e5274a27b2 000069/Assessing_gender_in_resilience_programming-Uganda_10215.pdf.

Pacini, G. C., Colucci, D., Baudron, F., Righi, E., Corbeels, M., Tittonell, P., and Stefanini, F. M. (2014). Combining multi-dimensional scaling and cluster analysis to describe the diversity of rural households. *Experimental Agriculture*, 50(3), 376–397.

Pitt, M. M., and Khandker, S. R. (1998). The impact of group-based credit programs on poor households in Bangladesh: Does the gender of participants matter? *Journal of Political Economy*, 106(5), 958–996.

Putnam, R. D. (1993). The prosperous community: Social capital and public life. *The American Prospect*, 4(13), 35–42.

Quisumbing, A. R., Brown, L. R., Feldstein, H. S., Haddad, L., and Peña, C. (1996). Women: The key to food security. *Food and Nutrition Bulletin*, 17(1), 1–2.

Rahman, R. A., and Dean, F. (2013). Challenges and solutions in Islamic microfinance. *Humanomics*, 29(4), 293–306.

Rahman, R. A., Al Smady, A., and Kazemian, S. (2015). Sustainability of Islamic microfinance institutions through community development. *International Business Research*, 8(6), 196.

Rao, N., Lawson, E. T., Raditloaneng, W. N., Solomon, D., and Angula, M. N. (2019). Gendered vulnerabilities to climate change: insights from the semi-arid regions of Africa and Asia. *Climate and Development*, 11(1), 14–26.

Renzhi, N., and Baek, Y. J. (2020). Can financial inclusion be an effective mitigation measure? evidence from panel data analysis of the environmental Kuznets curve. *Finance Research Letters*, 37, 101725.

Ritzer, G. (Ed.). (2007). *The Blackwell encyclopedia of sociology* (Vol. 1479). New York, NY, USA: Blackwell Publishing.

Robinson, M. S. (2001). The microfinance revolution: Sustainable finance for the poor. Washington, DC: The World Bank. Retrieved from: http://doc uments.worldbank.org/curated/en/226941468049448875/pdf/232500 v10REPLA18082134524501PUBLIC1.pdf.

Salami, A., Kamara, A. B., and Brixiova, Z. (2010). *Smallholder agriculture in East Africa: Trends, constraints and opportunities.* Tunis: African Development Bank.

Sarma, M. (2008). Index of financial inclusion. ICRIER Working Paper 215.

Sarma, M., and J. Pais. (2011). Financial inclusion and development. *Journal of International Development*, 23(5), 613–628.

Sarma, M. (2016). Measuring financial inclusion for Asian economies. In: *Financial inclusion in Asia* (pp. 3–34). London: Palgrave Macmillan.

Seman, J. A. (2016). Financial inclusion: The role of financial system and other determinants. Unpublished doctoral dissertation, University of Salford. Retrieved from: https://www.imf.org/en/Publications/WP/Issues/2016/12/31/Can-Islamic-Banking-Increase-Financial-Inclusion42710.

Sergado, C. (2005). Islamic microfinance and socially responsible investments. MEDA Project University of Torino. Retrieved from: http://www.gdrc.org/icm/islamic-microfinance.pdf.

Shahinpoor, N. (2009). The link between Islamic banking and microfinancing. *International Journal of Social Economics*, 36(10), 996–1007.

Shankar, S. (2013). Financial inclusion in India: Do microfinance institutions address access barriers. *ACRN Journal of Entrepreneurship Perspectives*, 2(1), 60–74.

Sharma, A. and Kukreja, S. (2013). An analytical study: Relevance of financial inclusion for developing nations. *International Journal of Engineering and Science*, 2(6), 15–20.

Sinclair, S. P. (2001). Financial exclusion: An introductory survey. Report of Centre for Research in Socially Inclusive Services, Heriot-Watt University, Edinburgh.

Sladkova, B. (2019). Insights from resilience policy work in Kenya: A Realist Evaluation Case Study. BRACED. Retrieved from: http://itad.com/wp-content/uploads/2019/11/Insights_from_resilience_policy_work_in_Kenya_realist_evaluation_case_study-1.pdf.

Smith, V. H. (2016). Producer insurance and risk management options for smallholder farmers. *The World Bank Research Observer*, 31(2), 271–289.

Stewart R., van Rooyen, C., Korth, M., Chereni, A., Rebelo Da Silva, N., and de Wet, T. (2012). Do micro-credit, micro-savings and micro-leasing serve as effective financial inclusion interventions enabling poor people, and especially women, to engage in meaningful economic opportunities in low- and middle-income countries? A systematic review of the evidence. Technical report, EPPI-Centre, Social Science Research Unit, University of London.

Sudmeier-Rieux, K. I. (2014). Resilience – An emerging paradigm of danger or of hope? *Disaster Prevention and Management*, 23(1), 67–80.

Summit, M. (1997). The microcredit summit report. Microcredit Summit. Retrieved from: http://www.microcreditsummit.org/resource/59/1997-microcredit-summit-report.html.

Suri, T., and Jack, W. (2016). The long-run poverty and gender impacts of mobile money. *Science*, 354(6317), 1288–1292.

Takahashi, K., Ikegami, M., Sheahan, M., and Barrett, C. B. (2016). Experimental evidence on the drivers of index-based livestock insurance demand in Southern Ethiopia. *World Development*, 78, 324–340.

The Economist. (2014). Poor service: Tiny Loans are getting more expensive. *The Economist.*

Tittonell, P., Muriuki, A., Shepherd, K. D., Mugendi, D., Kaizzi, K. C., Okeyo, J., Verchot, L., Coe, R., Vanlauwe, B. (2010). The diversity of rural livelihoods and their influence on soil fertility in agricultural systems of East Africa – A typology of smallholder farms. *Agricultural Systems*, 103(2), 83–97.

Tittonell, P., and Giller, K. E. (2013). When yield gaps are poverty traps: The paradigm of ecological intensification in African smallholder agriculture. *Field Crops Research*, 143, 76–90.

Tiwari, J., Schaub, E., and Sultana, N. (2019). Barriers to "last mile" financial inclusion: Cases from Northern Kenya. *Development in Practice*, 29(8), 988–1000.

Tura, H. T., Story, W. T., and Licoze, A. (2020). Community-based savings groups, women's agency, and maternal health service utilisation: Evidence from Mozambique. *Global Public Health*, 15(8), 1119–1129.

Umar, U. H. (2020). The business financial inclusion benefits from an Islamic point of view: A qualitative inquiry. *Islamic Economic Studies*, 28(1), 83–100.

USAID. (2012). Building resilience to recurrent crisis: USAID policy and program guidance. Washington, DC.

van de Steeg, J. A., Verburg, P. H., Baltenweck, I., and Staal, S. J. (2010). Characterization of the spatial distribution of farming systems in the Kenyan Highlands. *Applied Geography*, 30(2), 239–253.

Visser, W. A., and Macintosh, A. (1998). A short review of the historical critique of usury. *Accounting, Business and Financial History*, 8(2), 175–189.

Wairimu Ng'ang'a, T., and Crane, T. A. (2020). Social differentiation in climate change adaptation: One community, multiple pathways in transitioning Kenyan pastoralism. *Environmental Science and Policy*, 114, 478–485.

Weill, L. (2020). Islamic microfinance. In: *A research agenda for financial inclusion and microfinance* (pp. 99–110). Cheltenham: Edward Elgar Publishing Limited.

Weingärtner, L., Simonet, C. and Choptiany, J. (2019). Layering and tailoring financial services for resilience: Insights, opportunities and challenges of BRACED projects in Ethiopia, Kenya, Nepal and Senegal. BRACED Working Paper. London: Overseas Development Institute/BRACED.

Retrieved from: http://www.braced.org/resources/i/Layering-and-tailoring-financial-services-for-resilience/.
Wilkinson, E., and Peters, K. (2015). *Climate extremes and resilient poverty reduction: development designed with uncertainty in mind.* London: Overseas Development Institute.
Wilkinson, E., Budimir, M., Ahmed, A. K., and Ouma, G. (2015). *Climate information and services in BRACED countries.* BRACED Resilience Intel 1. London: Overseas Development Institute.
World Bank. (2018). *NEDI: Boosting shared prosperity for the North and North Eastern counties of Kenya.* Washington, DC: World Bank.
Xu, X. (2020). Trust and financial inclusion: A cross-country study. *Finance Research Letters*, 35, 101310.
Yoshino, N., and Morgan, P. J. (2017). Financial inclusion, regulation, and education: Asian perspectives. Tokyo: Asian Development Bank Institute. Retrieved from: https://www.think-asia.org/bitstream/handle/11540/8310/adbi-financial-inclusion-regulation-education-asian-perspectives.pdf?sequence=1
Zulkhibri, M. (2016). Financial inclusion, financial inclusion policy and Islamic finance. *Macroeconomics and Finance in Emerging Market Economies*, 9(3), 303–320.

CHAPTER 7

Disaster Risk Insurance

INTRODUCTION

Extreme weather events occur when a single local weather variable such as temperature or rainfall exceeds a particular threshold or when a combination of various variables occurs, as in the case of cyclones or droughts (Linnenluecke et al. 2012). They pose a major threat to fragile contexts and in particular, humanitarian settings where the impact of extreme weather events is leading to an increasing number of large-scale disasters. Climate and other natural hazard-related disasters that occur due to extreme weather events compound existing poverty, fragility and vulnerability, which undermine and threaten long-term development. While there is uncertainty about the impact of climate change-induced disasters in terms of the scope and timing of impacts, there is a growing concern that weather extremes such as heat waves, droughts, floods and hurricanes could be changing in frequency and intensity due to human influences (Trenberth et al. 2007).

Prior to an extreme weather event, there are often early warning signs available (several lead times before the event occurs) of heightened risks such as forecasts of storms, that indicate a high risk of flooding (Coughlan de Perez et al. 2015). These early warning signs provide a crucial window of opportunity to respond to and reduce both the potential impacts of an extreme weather event and disaster risks. Furthermore, there are several

© The Author(s), under exclusive license to Springer Nature
Switzerland AG 2021
M. Ahmed, *Innovative Humanitarian Financing*,
Palgrave Studies in Impact Finance,
https://doi.org/10.1007/978-3-030-83209-4_7

steps and preventative action that can be taken, such as an evacuation in the threat of flooding, which can reduce the negative impact of disasters. Taking advanced preparedness actions before a disaster occurs can be both effective in saving lives as well as reduce emergency response costs (Coughlan de Perez et al. 2016). Moreover, the vast majority of evaluations of preventative action reveal that avoided disaster losses can at least double or quadruple the investment in risk reduction (Mechler 2005).

In the current climate, where there is a significant rise of humanitarian disasters that are increasing needs, there has been an emergence of innovative risk financing mechanisms to improve the cost-effectiveness and timing of humanitarian funding. These instruments were designed to build more efficient national risk management systems to not only unlock critical funding needed to support better disaster responses, but to also help countries better prepare for and plan, for the impacts of climate change (Iyahen and Syroka 2018). Facilitating the access and use of these instruments could have the potential to enhance resilience by allowing humanitarian organizations to better prepare for, cope with and respond to disasters.

This chapter will evaluate the use and effectiveness of innovative risk financing mechanisms and is organized as follows; the next section will critically examine the relevant literature. Section three will assess the African Risk Capacity and in particular, will focus on the case of Malawi when the country experienced extreme weather events in 2015/2016, to evaluate the effectiveness of the risk financing mechanism and the lessons that can be learned from the Malawi case. Section four will conclude and provide further recommendations.

LITERATURE REVIEW

Humanitarian organizations were created to essentially respond to disasters after they had occurred as such, humanitarian interventions are inherently needs-based. Given the increase and frequency of crises, there is a growing need for humanitarian assistance to be forward-looking and risk-informed. It is estimated that only about 12% of funding in the past 20 years was invested in reducing the risk of disasters before they occurred while the rest of the funding was allocated toward emergency response, reconstruction and rehabilitation (Kellett and Caravani 2013). Also, there is a tendency for international aid to arrive several weeks after a disaster strikes (Awondo 2019). These inefficiencies coupled with the absence of early warning and ex ante mitigation systems for natural disasters, threaten food security and increase poverty (Zimmerman and Carter 2003; Kazianga and Udry 2006; Dercon and Christiaensen 2011). Action based on early warning systems has saved millions of lives and prevented significant damage where disaster managers have successfully used forecasts in cyclone-prone areas of the world (Galindo and Batta 2013; Harriman 2014; Lodree 2011; Rogers and Tsirkunov 2013). Furthermore, heatwave early warning systems, most commonly present in developed countries, can also trigger action to reduce mortality rates (Ebi et al. 2004; Fouillet et al. 2008; Knowlton et al. 2014). However, these systems sharply contrast with the lack of humanitarian action taken prior to predictable natural hazards such as flooding (Coughlan de Perez et al. 2016) where the barriers to early action are particularly evident in data-scarce areas in developing countries (Brown et al. 2007; Houghton-Carr and Fry 2006). One of the major barriers to this is the lack of funding that is available when a disaster is likely to occur but not certain (Coughlan de Perez et al. 2016) but this incentive structure is starting to change with the creation of new forecast-based financing systems (Coughlan de Perez et al. 2015).

Disaster Risk, Reduction and Management

By definition, disaster risk is a combination of hazard and vulnerability, with different methods being taken to join the two parameters depending on the theory or practice being used (Lewis 1999; Wisner et al. 2004, 2012). Furthermore, disaster risk reduction (DRR) is defined as the concept and practice of systematically managing the factors of disasters in order to reduce disaster risk (UNISDR 2016). Disaster risk management is "the application of disaster risk reduction policies and strategies to prevent new disaster risk, reduce existing disaster risk and manage residual risk, contributing to the strengthening of resilience and reduction of disaster losses" as defined by the UNDRR terminology (Albris et al. 2020). It includes anticipatory actions that work to minimize the creation of new risks and the reduction of any existent risks, in addition to the preparation for and response to disasters (IPCC 2012).

According to Szlafsztein (2020), among the actions that can reduce risks are emergency preparedness (such as planning and exercises, communication systems, public awareness and technical response capacity) and soft and hard infrastructure investments (such as strengthening and enforcing building codes and constructing defenses). Furthermore, during and after a disaster strikes, strategies and actions tend to be directed toward emergency response, the reconstruction and rehabilitation of damaged goods and economic and social recovery such as housing and transport (Miller and Keipi 2005).

Over the past few decades, there has been an increased need to acknowledge disaster risks in long-term development projects and plans, particularly after the signing of the United Nations-endorsed Hyogo Framework for Action in 2005 (Manyena 2012). Among the measures set out, the framework advocated for the use of financial risk-sharing mechanisms to reduce disaster risk and enhance climate change adaptation (Warner et al. 2009). Following on from the Hyogo Framework, in March 2015, the Sendai Framework for Disaster Risk Reduction (SFDRR) 2015–2030 was launched, it laid out a voluntary pathway for disaster risk reduction. SFDRR set out a list of goals and priorities to assess, develop and implement DRR strategies until 2030 (Gouramanis and MoralesRamirez 2020) and the four priorities for action SFDRR outlines are (UNGA 2015):

1. Understanding disaster risk
2. Strengthening disaster risk governance to manage disaster risk
3. Investing in disaster risk reduction for resilience
4. Enhancing disaster preparedness for effective response and to "Build Back Better" in recovery, rehabilitation and reconstruction

In regards to investing in disaster risk reduction for resilience, to achieve this outcome, the SFDRR aims to enhance local, national, regional and global capacity and capability to "promote mechanisms for disaster risk transfer and insurance, risk-sharing and retention and financial protection, as appropriate, for both public and private investment in order to reduce the financial impact of disasters on Governments and societies, in urban and rural areas" (UNGA 2015).

Disaster Risk Reduction and Climate Change Adaptation

Climate change adaptation (CCA) is defined by the IPCC as "an adjustment in natural or human systems in response to actual or expected climate stimuli or their effects, which moderates harm or exploits benefit opportunities" (Mercer 2010). Emergency management and DRR efforts are key to CCA (Nalau et al. 2016) as climate change is estimated to create more frequent and intense climate and weather fluctuations and in association with increases in exposure, will lead to greater destruction to human and environmental systems (Birkmann 2011; IPCC 2012).

According to Mall et al. (2019), when it comes to policy goals, both DRR and CCA have several similarities for instance, DRR is "concerned with a widely known problem in the field of all environmental hazards" and CCA "with only emerging issues related to climate change-instigated environmental hazards." Furthermore, DRR deals with all environmental hazards with the inclusion of CCA, which by definition is about climate change and the tools that are used to monitor, analyze and address adverse consequences (Kelman et al.). In addition to CCA, many countries have been making efforts to institutionalize several DRR-related practices in order to reduce vulnerability, however, many DRR measures particularly those related to events such as drought-proofing, flood protection, cyclone warning and shelters, malaria eradication, resistant agriculture, mangrove conservation, saline embankment and alternative livelihood development—are similar to CCA measure in regards to their application (Mall et al. 2019).

However, CCA and DRR have also been posed as somewhat contrasting issues and communities of practice (Gero et al. 2011; Schipper 2009; Thomalla et al. 2006; UNDP 2009; UNISDR and UNDP 2012), that are dedicated to similar aims (Ireland 2010; Schipper 2009). According to Nalau et al. (2016), differences are most apparent "in the ways key concepts and terms, such as resilience, vulnerability and adaptive capacity, are interpreted and used," which has led to differences in the way research, policy and practice are conducted (Ireland 2010; Moench 2009; Schipper 2009; UNDP 2009). Furthermore, the difference in their approaches is partly derived from the different underlying issues they are trying to address (Nalau et al.) as DRR institutions and policies were not designed for long-term strategic policy but rather for immediate responses while CCA on the other hand, emerged as a result of science and projections of potential effects from climate change (Handmer and Dovers 2013).

Linking DRR and CCA is seen as desirable as it could provide benefits at all scales through reducing vulnerability and by an increased focus on a multi-hazard approach (IPCC 2012) also, it could result in practical benefits such as increasing international funds for adaptation, more access to a broader range of expertise and by embedding a forward-looking approach in DRR by considering a longer timeline (Nalau et al. 2016).

Disaster Risk Financing and Anticipatory Action Approaches

In order to build disaster resilience, investment and financing is a key factor. According to Katongole (2020), disaster risk financing "is a proactive approach to disaster risk reduction which involves early detection of disaster risks and making financial resources available to take care of the needs of the affected communities during and after the disaster to smoothen consumption and engender resilience." It is a recent approach to DRR promoting financial resilience against disasters (Katongole 2020) and it also helps to safeguard economic growth and development when a disaster strikes. Moreover, disaster risk financing encompasses the financial aspects of a comprehensive disaster risk management system which includes both pre-disaster and post-disaster measures (Mita 2016; Juswanto and Nugroho 2017).

Governments use a variety of disaster risk financing instruments—such as accumulated reserves and savings, contingent credit, risk transfer through index-based insurance and re-insurance, post-disaster budget reallocations and post-disaster borrowing—whereby these mechanisms are used to fund disaster recovery efforts (Katongole 2020). There are various examples of disaster risk financing instruments such as cash transfers in Kenya, household reinsurance pools in Peru, disaster insurance in the Pacific Island countries and Catastrophe Deferred Drawdown Option in the Philippines—where all instruments are supported by the World Bank (Clarke and Wren-Lewis 2016). The aim of these risk financing instruments is to avoid governments from becoming emergency borrowers and to protect their fiscal balance (Katongole 2020).

While anticipatory action approaches are a wide range of initiatives that encompass a set of pre-planned measures that are taken before a disaster occurs, when a disaster is imminent or prior to acute impacts being felt from a disaster, these initiatives are anticipatory risk management mechanisms which include forecast-based early action, forecast-based financing and early warning early action. Given that 24.9 million people were displaced as a result of natural hazard-related disasters in 2019 (Internal Displacement Monitoring Centre 2020), there is a growing focus on the use of anticipatory action mechanisms to be put in place before a disaster occurs in order for there to be a rapid humanitarian response. Furthermore, anticipatory action approaches have been increasingly recognized by donors, humanitarian organizations and disaster risk managers as viable mechanisms to provide support to communities that are at risk, before disasters happen (Weingärtner and Wilkinson 2019). These anticipatory mechanisms are aimed at using forecasts or early warning signs of imminent disasters, in order to mitigate or reduce the impact of disasters as well as enhance post-disaster response.

Forecast-Based Financing

Forecast-based financing (FbF) refers to the use of weather forecasts and risk data to trigger funding for humanitarian action prior to a disaster event or before severe impacts are experienced by the at-risk populations (Coughlan de Perez et al. 2015). Moreover, it is an innovative instrument in the humanitarian sector that aims to reduce the impact of a disaster through anticipatory action, instead of waiting for a disaster to occur to initiate response and action (Coughlan de Perez et al. 2016). FbF was created out of institutional learning on the implementation of early actions taken based on early warnings and builds on an extensive body of work on early warning early action in the humanitarian sector (Alfieri et al. 2012; IFRC 2009; Krzysztofowicz 2001; Webster 2013).

In the past, forecasts have been successfully used to inform early warnings across a growing set of disasters (Gros et al. 2019). However, the extent to which anticipatory actions are taken varies according to the forecast's lead time and the degree of uncertainty, for example, storms and cyclones are relatively short-term risks that can be forecasted relatively well, hence, if there is an imminent threat of a cyclone occurring, early action can be taken in cyclone-prone areas to prevent significant damage and mitigate the impact of the disaster (Galindo and Batta 2013; Lodree 2011; Rogers and Tsirkunov 2013; Harriman 2014). A case in point is India where before cyclone Phailin hit in 2013, it is estimated that up to 800,000 people were evacuated within 48 hours before the cyclone happened based on weather forecasts (Ghosh et al. 2014).

However, in the context of longer-term risks or when forecasts have a high level of uncertainty such as in the case of floods, these large-scale anticipatory measures are less feasible, particularly in data-scarce areas (Gros et al. 2019). Nevertheless, in such cases with a seasonal lead time, there are a number of risk reduction actions that can be taken to avert losses. For example, before flooding occurred in West Africa in 2008, the International Federation of Red Cross and Red Crescent (IFRC) was able to take anticipatory action and source disaster management supplies ahead of time based on seasonal forecasts of above-normal rainfall, this anticipatory action resulted in the improvement of the disaster management supply availability from 40 days to two days, to when flooding did occur (Braman et al. 2013). Furthermore, in East Africa, the IFRC has used severe-weather warnings of above-normal rainfall to reduce the risk of diarrheal disease outbreaks by organizing the maintenance of drainage channels, repairing latrines, storing food and water and gathering supplies for boiling water (Red Cross Red Crescent Climate Centre 2013).

In Bangladesh, investments in flood forecasting have enabled strong early warning systems to anticipate cautions for at-risk communities in any given event, including impact-based forecasting and flood damage functions (Yang et al. 2015; Sai et al. 2018; Priya et al. 2017; Webster et al. 2010). Such early warning systems are a key mechanism to reduce disaster impacts and address the increasing number of climate-induced risks (Watts et al. 2018). This is especially significant given that due to its geophysical settings, Bangladesh is highly vulnerable to climate change-induced catastrophes as globally, Bangladesh is one of the most disaster-prone countries, with flooding an annual regularly occurring event and during the monsoon season, typically 80% of the annual amount of rain falls (Gros et al. 2019). Bangladesh was therefore one of the first countries in the world to operationalize FbF in the anticipation of a natural disaster as a result of flooding and the first to implement forecast-based cash distribution in such circumstances (Gros et al. 2019).

FbF as a financing instrument enables the automated release of funds in the anticipation of an extreme weather event through Early Action Protocol (EAP) to implement pre-defined actions (Gros et al. 2019). Moreover, in the EAP, involved stakeholders pre-decide on the forecast-based triggers that initiate them, on a set of actions and how the actions are going to be implemented and by which organizations (Coughlan de Perez et al. 2015). Therefore, once the likelihood of a forecasted disaster of a certain magnitude is above a specified threshold of probability, decisions are automatic which therefore allow institutional actors to make the most efficient use of limited lead times, avoid making hasty decisions and enable more efficient planning with sufficient stakeholder consultation (Gros et al. 2019). This is significant given that previously, a significant barrier to effective early action was the ability to release humanitarian funds based on forecasts.

Insurance Mechanisms

An important element of anticipatory financing is the use of insurance which has been most commonly used in relation to natural hazard-related risks (such as droughts), as the economic case for proactive management of disaster risks in order to avoid further risks and devastation is strong (Surminski and Tanner 2016). Insurance is a mechanism for funding extreme and non-gradual climate events recovery such as floods, windstorms and droughts, but it is not suited to manage disaster caused by slow-moving or gradual changes such as sea-level rise, desertification, loss of habitat, loss of biodiversity, erosion, ocean acidification and glacial retreat, among other impact (Bouwer 2019).

Risk reduction strategies and forward-looking climate adaptation are important to allow individuals, businesses and governments to build resilience against the impacts of extreme weather events and long-term changes (Surminski et al. 2019). There is growing interest from governments, donors, businesses and civil society to use financial risk transfer as an intervention tool and resilience measure (Surminski et al. 2016). Insurance is considered as a possible tool to reduce or compensate for economic losses from disasters through ex ante risk management. According to Linnerooth-Bayer et al. (2003), insurance-related instruments that spread and pool risks may be important tools for supporting adaptation to climate-related disasters in developing countries.

Traditional risk financing mechanisms, such as insurance, help households to survive shocks to their livelihoods and assets when a disaster strikes by pooling risks across communities and regions (Linnerooth-Bayer and Hochrainer-Stigler 2015). In the absence of insurance or government assistance, households that have been impacted by a natural disaster may have to sell productive assets. For instance, during the 1984–1985 drought-induced famine in Ethiopia, many households lacked adequate government assistance and had to sell productive assets, consequently during the 1990s, those same households continued to experience considerably less per capita annual growth than households who had not experienced the drought (Wiseman and Hess 2007).

In regards to donors, insurance helps them to reduce post-disaster liabilities (Linnerooth-Bayer et al. 2005). However, in low-income countries and in rural areas in particular, formal insurance markets are often incomplete and non-existent. According to Barnett et al. (2008), a common reason for insurance market failure in low-income countries is the lack of effective legal systems to enforce insurance contracts, furthermore if there are effective contract enforcement measures in place, insurance markets often fail due to other reasons such as asymmetric information issues and high transaction costs.

Index-Based Micro-insurance

An innovative risk financing insurance mechanism is index-based micro-insurance. The aim of micro-insurance is to circumvent the high costs associated with traditional insurance by offering limited cover thereby significantly reducing transaction costs in order to serve low-income markets (Hochrainer-Stigler et al. 2012). Micro-insurance can be indemnity-based whereby products are written against losses that have incurred, or they can be index-based, where products are written against physical or economic triggers, in other words, they can be written against events that cause actual loss and not against the loss itself (Linnerooth-Bayer and Hochrainer-Stigler 2015). Index insurance therefore reduces the issue of moral hazard as claims are independent of losses (Linnerooth-Bayer and Hochrainer-Stigler 2015). Index-based micro-insurance tools have also been created to provide protection against the loss of assets and investments. For these types of mechanisms, payouts are determined according to a pre-agreed index such rainfall levels or yield data in order for funds to be disbursed. Examples include agricultural insurance which has been used by several countries as a safety net in order to protect farmers and combat food security concerns (Golnaraghi et al. 2016; Tanner et al. 2015).

Insurance mechanisms also have additional benefits such as enabling productive investments and therefore helping those that are of high-risk avoid disaster-induced poverty traps (Barnett et al. 2008). In the case of a drought, loss of herds can have a devastating impact on local households and may push them further into dire poverty. To potentially mitigate against this, index-based livestock insurance has been piloted in Ethiopia to help stabilize asset accumulation and enhance economic growth by insuring livestock keepers against the loss of their livestock if a drought were to occur in order to help keep them out of poverty traps (Greatrex et al. 2015). The private sector has taken an interest in micro-insurance markets with insurers such as Swiss Re targeting markets including those who can afford commercially viable premiums which they estimate to be 2.6 billion people living above the international poverty line of USD$1.25/day but below USD$4/day (Linnerooth-Bayer and Hochrainer-Stigler 2015; Swiss Re 2012). However, there are relative few insurers who are optimistic about the prospects of disaster micro-insurance for the extreme poor (those living below USD$1.25/day) unless it is supported by governments, non-governmental organizations or foreign donors (Linnerooth-Bayer and Hochrainer-Stigler 2015).

Sovereign Disaster Insurance

Post-disaster, if governments lack the capital needed to provide human-itarian assistance and restore and rebuild critical infrastructure, then indirect costs can exceed the direct losses of a disaster (Global Facility for Disaster Reduction and Recovery 2013). Furthermore, the lack of government support may lead to severe secondary economic and social consequences at the national level such as an increase in poverty, budget imbalances and declines in trade (Barnett et al. 2008). For example, following the tropical storms that hit Yemen in 2008, damages and losses of approximately USD$1.6 billion were caused which roughly estimated to be 6% of the country's GDP (Linnerooth-Bayer and Hochrainer-Stigler 2015) also, in some regions, poverty rates increased from 28 to 51% and the national poverty level increased by 1.1% as a result (Global Facility for Disaster Reduction and Recovery 2009). For small countries or coun-tries that are highly exposed to disasters, insurance and other risk transfer instruments should be considered if they are unable to rely on a tax base to raise the funding needed following a disaster (Linnerooth-Bayer et al. 2005; Hochrainer and Pflug 2009). Moreover, risk financing through tools such as insurance mechanisms can be used in anticipation of a catastrophe impacting an area or a region. Governments can use these insurance instruments, such as sovereign disaster risk insurance, to spread risks and to help them manage disasters such as earthquakes, droughts, floods or hurricanes (Martinez-Diaz et al. 2019).

There are a number of innovative sovereign disaster insurance mech-anisms that have been used in developing countries. For instance, in Turkey, the first national public–private partnership in the developing world that was created to provide affordable cover for earthquake risk in urban areas was known as the Turkish Catastrophe Insurance Pool (TCIP) (Linnerooth-Bayer and Hochrainer-Stigler 2015). The purpose of the TCIP was to reduce the Turkish government's fiscal exposure by building up capital overtime in an insurance pool that is funded by mandatory private contributions and guaranteed by the government as well as donors (Gurenko 2004). The insurance pool was created as a way to circumvent the issue of private insurers reluctant to offer regional or nation-wide policies to countries to cover high risk-level events such as droughts, floods and other hazards because of the dependent or co-variant kind of risks involved (Linnerooth-Bayer and Hochrainer-Stigler 2015). In some cases, sovereign disaster risk insurance mechanisms operate through regional risk pools and are parametric in nature meaning if, for example, there are low levels of rainfall, ex ante agreements secure the release of future funds based on these types of triggers.

Regional Insurance Pools

In post-disaster scenarios, the need for critical and rapid expenditure can lead governments to use slow or expensive instruments such as budget reallocations or borrowing with unfavorable terms (Benson and Clay 2004). As the severity and frequency of climate-induced extreme weather events continue to rise, governments are needing to more so consider alternative ways of meeting the financial consequences of such events. To address this and in an attempt to better financially prepare for disasters, there has been an increased interest in implementing comprehensive sovereign disaster risk finance strategies of governments which are defined by the World Bank as "the bringing together of pre- and post-disaster financing instruments that address the evolving need of funds – from emergency response to long-term reconstruction – and are appropriate to the relative probability of events" (Clarke et al. 2017). Furthermore, innovative sovereign risk financing mechanisms such as regional pools may help to increase the financial resilience of the corresponding risk bearers (Linnerooth-Bayer and Hochrainer-Stigler 2015).

Developing countries, and small island states in particular, tend to pay international prices for insurance which is subject to fluctuations caused elsewhere (Linnerooth-Bayer and Hochrainer-Stigler 2015). For example, after Hurricane Andrew hit in 1992, Barbados experienced a ten-fold increase in insurance premiums, this is despite the fact that the country was not impacted by the Hurricane or lying in a major hurricane path (Cummins and Mahul 2008). For larger countries, the impact of a severe natural disaster can be absorbed as an impacted region can be subsidized by revenues from unaffected regions, however, for small island states, this type of geographic risk distribution is not possible (Linnerooth-Bayer and

Hochrainer-Stigler 2015). Most Caribbean countries are highly exposed to adverse natural disasters which can have devastating economic impacts. It is estimated that on average, there is at least one major hurricane and numerous tropical storms to hit the Caribbean each year but due to their small size, Caribbean states often have limited financial capacity to respond to natural disasters (Ghesquiere and Mahul 2007).

To address this issue, the Caribbean Catastrophe Risk Insurance Facility (CCRIF) was created to provide governments with immediate access to liquidity in the event of a major disaster, such as a hurricane or earthquake, at a significantly lower cost than if governments in the Caribbean were to purchase insurance individually on the global financial markets (World Bank 2007). The CCRIF was the first multi-country risk pool to be established and for the CCRIF, governments of 16 Caribbean island states contribute resources to the pool which is contingent on their risk, and the fund is reinsured in the capital markets (Linnerooth-Bayer and Hochrainer-Stigler 2015). Subsequently, an insurance pool of Pacific island countries was established by the World Bank, with the support of the Government of Japan, following the CCRIF model (Linnerooth-Bayer and Hochrainer-Stigler 2015) where five of the Pacific island states received protection against disasters such as earthquakes, tsunamis and tropical cyclones from Swiss Re and other insurers at a rate lower than if the states insured themselves separately (Pacific Catastrophe Risk Assessment and Financing Initiative 2013). Similarly, the African Risk Capacity was established as a continent-wide mechanism for pooling risk following similar principles of the CCRIF model (Lucas 2015).

CASE STUDY: AFRICAN RISK CAPACITY

Over the last few decades, extreme weather-related losses have been on the rise (Awondo 2019). In 1981, 2001 and 2014, extreme weather-related losses accounted for USD$16 billion, USD$33 billion and USD$95 billion, respectively (Munich Re 2015; The World Bank 2013), where developing countries, with predominantly agrarian economies that are rain-fed and with underdeveloped financial and insurance markets, being the most vulnerable (Awondo 2019).

Mechanisms to fund recovery efforts range from government assistance, humanitarian aid, savings and credit, informal risk sharing to insurance and alternative risk-transfer mechanisms (Linnerooth-Bayer and Hochrainer-Stigler 2015). While international assistance may be of help after a disaster, voluntary donations from international NGOs, governments and individuals, have only averaged an estimated 3% of direct economic losses in developing countries, although the amount is significantly higher for widely publicized events (Becerra et al. 2014). For catastrophic high-risk events, disaster victims in fragile countries typically rely on their government and donor assistance to provide relief and for small and highly exposed countries, governments can face severe post-disaster fiscal constraints which will limit their capacity to backup private insurers or cover their post-disaster costs to help restore public infrastructure (Linnerooth-Bayer and Hochrainer-Stigler 2015).

Nine of the top ten countries that were most impacted by extreme weather events and losses between 1996 and 2015 were in developing countries in addition, two of the top three countries most affected in 2015 were in the African continent, according to the Global Climate Risk Index (Kreft et al. 2017). Over the years, African countries have experienced frequent droughts which can more than often threaten food security and livelihoods (Kehinde 2014) and in the event of a disaster, Africa relies mainly on international aid assistance for managing risks ex-post, for example, in 2012, USD$2.7 billion was spent in the form of international aid by the World Food Program in Africa which was 66% of their global expenditure (Awondo 2019). In an effort to better manage extreme weather-related risks, in November 2012, the African Union established the African Risk Capacity to pool extreme risks across the African continent, such as droughts, floods, earthquakes and cyclones, using insurance and reinsurance from international financial markets (Awondo 2019).

Background on the African Risk Capacity

The African Risk Capacity (ARC) is an agency that was established by the African Union in November 2012 and is comprised of two entities (1) the Specialized Agency and (2) the ARC Insurance Company Limited, a financial affiliate company. The objective of ARC was to help African countries cope with the effects of drought as Africa, being mostly an agricultural-based economy, relies on adequate rainfall for sufficient crop yield (Kehinde 2014). In general, the agency provides oversight and supervision of the development of ARC capacity and services including the capacity building of individual countries, approving contingency plans and monitoring their implementation (Awondo 2019).

ARC was established to link insurance payout to effective response and contingency plans (Kehinde 2014), furthermore, ARC carries out commercial insurance functions such as risk pooling and risk transfer with the principal aim of harnessing the natural diversification of weather risk across Africa, thereby allowing countries to better manage their risks collectively in a timely manner as well as efficiently, at a reduced cost in an effort to prevent humanitarian crises (Awondo 2019). The aim of ARC was "to help African governments reduce the negative impact of droughts on the lives and livelihoods of the vulnerable, while decreasing reliance on external aid" (Kehinde 2014). In the case of extreme drought events, based on objective triggers, ARC aims to give countries access to immediate funds thus reduce dependence on international appeals for emergency food aid assistance (Clarke and Hill 2013).

For ARC membership, national governments collaborate with ARC to develop an Operation Plan and a Final Implementation Plan that is to be submitted shortly before a payout, which provides details on how ARC payout is to be deployed in a given scenario (Awondo 2019). Furthermore, upon completion, member countries pay a one-off contribution to join the insurance pool subsequently, ARC works with member states to calculate country-wide premiums and allocate payouts based on prearranged and transparent rules for payment (Awondo 2019). Each government has to develop a contingency plan for how they intend to use any claim payments and there are restrictions on how governments can distribute funds (Clarke and Hill 2013). Moreover, each member country chooses their coverage level and payouts, and ARC payouts are targeted to reach national treasuries between two and four weeks after harvest (Awondo 2019).

The initial phase of the project was limited to developing a region-wide rainfall index insurance in order to pool extreme drought risk across a pilot number of African countries and the first financial affiliate of ARC (ARC Ltd) was created at the end of 2013 as a mutual insurer (Awondo 2019). In May 2014, ARC Ltd issued its first weather (rainfall) index insurance policies, that was triggered by a drought, to a first of its kind risk pool consisting of Kenya, Mauritania, Niger and Senegal, with the support of the government of Germany, amounting to USD$130 million in liabilities and USD$17 million in annual premiums (Awondo 2019). The agency then made its first payout in 2014/2015 for drought-related losses to Mauritania, Niger and Senegal for just over USD$26 million and in 2015, an additional five countries joined the pool (Burkina Faso, The

Gambia, Malawi, Mali and Zimbabwe) which resulted in a total liability of over US$190 million in drought coverage (Awondo 2019). In regards to the premiums for each member state upon establishment of ARC, the majority of the premiums were expected to be paid for by foreign donors, at least in the medium term (Clarke and Hill 2013). For each member state, the premiums were 75% financed by the Government of Germany during the first year and 50% financed during year two and 25% financed during year three and by the fourth year, each country was expected to be responsible for 100% of the premiums (Awondo 2019).

How the African Risk Capacity Works

The ARC is based on two key components, African Risk View (ARV) software as well as an index-based contingency funding pool (Lung 2013). ARV is ARC's modeling platform that provides modeling input to ARC for insurance purposes, furthermore, it also aims to be an early warning financial tool that can potentially trigger early action and risk reduction measures (LinneroothBayer et al.). In particular, ARV is a drought impact modeling platform aiming to quantify and monitor weather-related food security risk in Africa (Calmanti et al. 2012) and based on rainfall data, it calculates the number of people in need of food assistance in a given African country in times of drought (Lung 2013). By using climate change scenarios as an input to ARV, in theory, it should be possible to evaluate the future impact of climate variability on critical concerns such as food security and the overall performance of the predicted risk management system (Calmanti et al. 2012).

ARV operates by comparing actual rainfall data with a previously derived benchmark value then predicts at the time of harvest how many people are in need of food assistance and how costly an intervention would be (Lung 2013). It is anticipatory in nature as it determines beneficiary data and intervention costs prior to spending. Previously, these figures were determined by the WFP through lengthy damage assessments conducted only after farmers began to experience the impacts of drought, therefore by replacing the damage assessment with a calculation that estimates the extent of the impacts of drought prior to it being felt, should in theory allow for an earlier humanitarian response (Lung 2013) and potentially the offsetting of longer-term damage.

Pay-out decisions are made using ARV which uses satellite-based rainfall data to estimate if the water requirements of a certain crop for a given country have been satisfied (Reeves 2017). When these are not met, it uses static information about population vulnerability in order to estimate the number of people impacted by the shortfall, it then converts this into a response cost (Reeves 2017). Payouts are triggered from ARC to countries if the estimated response cost at the end of the season exceeds a pre-defined threshold as agreed upon in the insurance contracts (ARC 2016). As an indicator for drought, ARV uses the Water Requirements Satisfaction Index (WRSI), an index which calculates if a particular crop is receiving the amount of water it requires at different stages of its development, which was developed by the Food and Agriculture Organization (FAO) of the United Nations (ARC 2016). Moreover, the WRSI is an indicator of the performances of agriculture and uses historical records of food assistance operations in order to estimate future potential needs for livelihood protection (Calmanti et al. 2012).

ARC's parametric insurance products are reliant upon the adequate functioning of ARV, the model must therefore provide an accurate enough assessment of the actual impact of drought on food security in order to be trusted by countries who would like to purchase parametric weather insurance (Scott et al. 2017). Although it can be argued that this trust has been achieved in some cases, there is a lot of skepticism with the model given the recent experience and somewhat failure of ARV in Malawi (Scott et al. 2017).

Malawi Country Context

It is predicted that countries in Southern Africa will be significantly impacted by increasing climate change and variability (Lobell et al. 2008). Furthermore, key factors affecting food systems in most areas in sub-Saharan Africa are climate change and weather variability (Kotir 2011), particularly among rural subsistence farming communities (Shisanya and Mafongoya 2016). In low-income countries, where the population predominantly depends on subsistence farming, climate change is an ongoing concern (Suckall et al. 2017) and with the rapid changes in climate anticipated to continue to negatively impact agricultural production across Africa, small-scale farmers are particularly more vulnerable (Easterling et al. 2007), resulting in widespread poverty and food insecurity (Nkomwa et al. 2014).

With sub-Sahara Africa being one of the most vulnerable regions to climate change (Joshua et al. 2016), Malawi has been experiencing increases in extreme weather events, such as floods and prolonged dry spells, that is significantly impacting the population and the economy (McSweeney et al. 2010). In addition, the country has also been experiencing other climate change-induced issues such as strong winds and heatwaves (Mutanga et al. 2012) that have been impacting energy access, water and food security (Khamis 2006).

Malawi is a country located in South-East Africa and is one of the poorest countries in the, world ranking 171 out of 188 countries in the 2018 Human Development Index (UNICEF 2019) and by 2018, Malawi had a population of 17.5 million people, the majority of which were living in rural areas (National Statistical Office 2019). With ~150 people per square kilometer in the southern regions and over 50% of rural households occupying <1 ha of land—Malawi has one of the highest population densities in Africa (National Statistical Office 2015; FAOSTAT 2017). Poverty impacts an estimated 59.5% of the rural population (National Statistics Office 2017) and there is high dependency on low-productivity small-scale farming, which is considered the biggest barrier to rural poverty reduction (UNDP 2016). Moreover, in the rural areas of the country, the poverty levels are particularly high where people have no access to basic services such as water, sanitation and electricity among other services (Chamdimba et al. 2020).

Malawi has a sub-tropical climate with one crop growing season that tends to run from December to April (Mutegi et al. 2015). 90% of the labor force is located in rural areas and the agricultural sector is one of the biggest sectors that contributes an estimated 33% to the country's Gross Domestic Product (GDP) and 80% to national employment (National Statistical Office 2019). Nevertheless, the sector is highly vulnerable to the impacts of climate change and is highly reliant on rain-fed agriculture (Nkomwa et al. 2014). Moreover, it is estimated that more than 84% of the population in Malawi depend on natural resources-based livelihoods and rain-fed agriculture (ActionAid 2017) which are impacted by climate change.

In Malawi, periodic droughts and floods have become more frequent and there is an increased urgency to find viable solutions for farmers (Tesfaye et al. 2015) as agriculture is one of the biggest sectors and is largely rain-fed, practiced by 76% of Malawi's population (National Statistics Office 2017). For farmers who are reliant on the natural environment, increased climate variability, chronic environmental stresses and acute climate change-associated shocks threaten livelihoods, food security and overall wellbeing (Suckall et al. 2017). There is a plethora of adaptation-focused literature on how rural farmers can cope with climate change (Adger et al. 2003; Deressa et al. 2009; Stringer et al. 2009). However, there are concerns that the adaptation strategies suggested will not be sufficient in regions such as those in sub-Saharan Africa where high temperatures, inconsistent rainfall, land degradation, droughts and flooding are likely to have significant environmental effects and significant societal consequences (IPCC 2007; Conway 2009; Parry et al. 2007).

2015/2016 Malawi Extreme Weather Events

Malawi is a country that has an extensive history of experiencing recurrent famines induced by drought and flood (Haug and Wold 2017). From 1967 to 2014, Malawi suffered a total of seven severe droughts and 19 floods that adversely impacted smallholders' production and food security (Government of Malawi 2015). Furthermore, in the 1990s and the first half of the 2000s, severe drought-induced hunger caused Malawi to be at the forefront of international crisis headlines (Haug and Wold 2017). However, from 2005 to 2015, Malawi was able to transform itself from being a large humanitarian aid recipient to becoming self-sufficient in staple food and being an exporter of maize to Zimbabwe and Kenya (National Statistics Office 2015b; Government of Malawi 2016), in addition to the country's more traditional agricultural export of tobacco (Haug and Wold 2017).

However, in 2015, Malawi received the highest record of rainfall for the country causing severe flooding throughout the country and in particular the Southern Region (Government of Malawi 2015) and in 2016, another state of emergency was declared in the country with severe drought caused by El Niño, which left half of the population in Malawi severely food insecure and in need of food relief (WFP 2016). 2.8 million people were impacted by floods in 2015 and in 2016, 6.5 million people were impacted by drought as recorded by the Malawi Vulnerability Assessment Committee (Malawi Government 2015a, 2016). Overall, the estimated costs of flood damage were US$335 million, and the recovery plan was estimated at US$494 million which resulted in a decrease of 0.6% in annual GDP growth (Malawi Government 2015b).

These events led to a 30.2% year-on-year drop in maize production (World Bank 2019) which is grown on 90% of cultivated land (Mutegi et al. 2015) and the reduction in maize yields in 2016 led to 6.5 million people in Malawi needing food aid (USAID 2016). The Minister of Agriculture told the press: "Prices for maize, the nation's staple crop, have in recent months gone up more than 60% above the 3-year average for this time of the year, making it increasingly difficult for many people to buy food," the Minister also said that an estimated 1.2 million tons of maize would be needed to prevent the growing hunger crisis (Banda 2016).

According to predicted future climate change scenarios, not only are El Niño events predicted to become more extreme (Wang et al. 2019), but extreme weather conditions such as erratic rainfall, droughts and flooding are expected to become more frequent and the normal (Niang et al. 2014; Mittal et al. 2017; Hart et al. 2018). Risks of drought and flooding are prompting an increased focus on climate adaptation and emergency responses that take into consideration long-term resilience (Arndt et al. 2014; Challinor et al. 2016).

Malawi's ARV Failure

Following the 2015/2016 extreme weather events in Malawi, ARC came under scrutiny and was heavily criticized for the lack of immediate payout (Scott et al. 2017). According to a statement issued by ARC (2017),Malawi had purchased an insurance policy from ARC for its 2015/2016 crop season, at a time during which the country faced severe drought and although ARC funding was designed to be delivered swiftly in such cases, a payout was not immediately triggered. Instead, there was an agreement to a payout to Malawi in November 2016 of USD$8 million (several months after a national emergency was declared in April 2016) but the payment was only made in January 2017 (Reeves 2017). In the meantime, the Government of Malawi was left to explore other conventional means of raising capital to tackle the food insecurity issue with the total drought response costs estimated at USD$395 million (Reeves 2017).

Critics argued the insurance policy for which Malawi had paid USD$4.7 million for, failed to provide timely assistance to 6.7 million food insecure Malawians due to some major defects in the model (Reeves 2017). On the one hand, the Malawi incident has been used as case study to demonstrate the incompetence of market-based insurance products to respond to climate change-related food insecurity issues (Scott et al. 2017), in particular the insurance model itself as a mechanism to address climate change-induced risk, especially drought risk, has been criticized (Reeves 2017). While on the other hand, it has been argued that the response taken by ARC demonstrates the flexibility and willingness to identify and address the cause of the issue (Scott et al. 2017).

ARV failed to anticipate the impact of the drought on food security (Scott et al. 2017), moreover for Malawi, the ARV model indicated far lower numbers of people impacted by the drought compared to the actual impact of the drought on the ground (ARC 2017). ARC's main argument for the model's failure to predict the impact of the drought is, "the longer maturation period used in the model estimated a less severe impact of the dry spells. On the other hand, shorter maturing varieties increasingly planted by farmers were severely affected by a three-week dry spell that hit during flowering and that dramatically affected production (thus the failure of ARV to trigger a payout)" (Scott et al. 2017).

Lessons Learned

There are several lessons to be learned from the Malawi case.

1. **Timing issues/delayed payment**
   First, when evaluating ARC as a financing mechanism, it is important to take timing into consideration as there was a significant delay and ARC payout did not arrive in a timely manner. One of the strongest arguments for ARC products is "the relative speed with which a country would receive payout funds compared to other sources" (Scott et al. 2017). However, as demonstrated through the Malawi case, ARC payouts will not necessarily be triggered or funds mobilized fast enough to provide a timely response as when the threshold is reached for triggering a payout, the humanitarian situation would already be severe for a large segment of the population (Scott et al. 2017). ARC products should therefore be considered among other financing mechanisms that can provide similarly rapid access to funds (Scott et al. 2017).

2. **Insurance mechanisms are not a panacea**
   Second, designing innovative products that address drought impact (ex post) through the adoption of suitable ex ante strategies is key to forming holistic policies for reducing vulnerability to drought, in particular to smallholders (Tsegai Kaushik). However, ARC was designed to provide a response of last resort in cases of severe

drought, rather than function as a savings mechanism (Scott et al. 2017). In other words, insurance mechanisms are not a quick fix for the humanitarian financing system as demonstrated by the Malawi case (Reeves 2017) as according to the ARC (2017), its product are "not designed to cover every aspect of a Member State's response to natural disasters and weather events or the entirety of the costs involved." Instead, it should be part of a suite of services ARC members should rely on to help them be fully prepared as "it is not intended to be, nor does it claim to be, a replacement for large-scale international humanitarian intervention" (ARC 2017). Therefore, a key lesson that can be learned from the Malawi case is sovereign risk pooling mechanisms such as the ARC is not expected to plug the shortfall of funding after a large-scale humanitarian crisis occurs such as drought-induced famine. For instance, if Malawi had received the payout when it expected it to in 2016 rather than 2017, the sum would still had been relatively small compared to the total costs incurred due to the drought, therefore as a funding mechanisms, ARC is not considered to be solely capable of post-disaster financing.

3. **Other factors involved that can limit the role of insurance mechanisms**
   Third, with early warning analyses, two types of errors occur, they can either provide an overestimate or provide an underestimate of the severity of food insecurity and there is a "missed crisis" error, whereby the latter category occurs when forecasts miss crises or deteriorations of food security conditions (Krishnamurthy et al. 2020). In either case (overestimation or underestimation), there are other factors involved such as the uncertainties associated with seasonal weather forecasts, political instability or other factors (Ahmadalipour and Moradkhani 2017; Thiboult et al. 2017) and the costs can be significant particularly to food security issues. Therefore, it is too simplistic to attribute food insecurity issues solely to rainfall as there are other factors that impact food security projections such as sudden conflict, unexpected market conditions or pest outbreaks (Sandstrom and Juhola 2017).

In the experience of the extreme weather events in Malawi, in 2015/2016, a significant failure to trigger insurance payout was due to an incorrect agrifood system model (Reeves 2017). Although maize was selected as the reference crop for ARV, farmers were in fact growing a different/hybrid variety of maize in significant and increasing amounts also, there was out of date information on farming practices that "prevented the model from accurately replicating conditions on the ground at the end of the season" (ARC 2017). This meant that incorrect information was used for ARV resulting in inaccurate calculations and therefore an underestimation of the severity of food insecurity.

Conclusion

In summary, to better manage extreme weather risk in sub-Saharan Africa, ARC was established by the African Union to pool extreme weather risk across its member countries. As for many governments on the continent, the increasing risk from climate-induced events coupled with an overwrought humanitarian aid system struggling to respond to increasing needs, led to an uptake of innovative risk financing mechanisms such as ARC, to help governments access critical capital needed to respond to disasters in an efficient and timely manner.

This chapter examined the case of Malawi when in 2015/2016, extreme weather events did not result in ARC payouts being immediately triggered (as they were designed to) in order to help the government of Malawi respond early to the growing humanitarian crisis due to defects in the model, among other factors. Instead, Malawi received payouts several months after a national emergency was declared, while in the meantime, the drought left millions of Malawians food insecure. Although innovative risk financing mechanisms such as ARC have been mainly used by governments, there has been an increased interest from humanitarian organizations to develop and use similar risk financing instruments. However, for humanitarian organizations that are interested in similar insurance mechanisms to help them respond to climate-related disasters in humanitarian contexts, the case of Malawi is a caution and offers several valuable lessons.

Further Recommendations

1. **Insurance should not be seen as an all-encompassing funding instrument to respond to climate-related disasters in humanitarian contexts**
   Insurance mechanisms cannot provide full financial protection against the adverse effects of climate change (Warner et al. 2009)

as demonstrated in the Malawi case. Insurance mechanisms therefore have clear limitations, as in theory, they can only cover events that when they occur, they are "sufficiently random and infrequent" (Linnerooth-Bayer et al. 2019).

ARC payouts are supposed to use forecast-based methods to trigger funding for humanitarian response before climate-related events become acute, however, going forward, as climate-related humanitarian crises are set to intensify, insurance mechanisms such as ARC are not likely to provide governments with the financial security needed to cope with natural disasters. As demonstrated by the Malawi case, governments will still need to rely on humanitarian organizations and foreign donors to provide humanitarian assistance in order to deal with natural disasters and climate-induced issues such as food insecurity therefore, sovereign insurance mechanisms such as ARC, should be clearly presented as such and should not be considered to be a panacea.

2. **Linking insurance mechanisms with other financing mechanisms**
   In the case of the extreme weather events that took place in Malawi 2015/2016, the case has shown that insurance mechanisms such as ARC should not be considered as stand-alone funding solutions. Instead, how sovereign risk insurance instruments such as ARC can be linked to other disaster risk financing instruments in order to better manage climate-related disasters and risks, needs to be considered. In particular, designing innovative products that link insurance mechanisms with other disaster risk funding instruments should be explored and piloted. Furthermore, designing products that integrate DRR and CAA could potentially be helpful in addressing ex post drought impact, in addition to implementing ex ante strategies, in order to help reduce vulnerability to drought in the future.

3. **Modify ARV modeling to incorporate changes in farming practices and reference crops**
   As mentioned above, one of the main failures of ARV was the fact that maize was chosen as the reference crop yet farmers in Malawi were in fact growing a hybrid variety of maize which lead to incorrect information for ARV calculations. Although ARV could be considered as functioning well, it will not be effective in fulfilling its aim in triggering ARC payouts to provide a timely response or be an effective decision-making tool, if the correct data is not incorporated into ARV. Going forward, similar incorrect information could be used by ARV if on the ground changes in farming practices or changes to reference crops are not accurately reflected.

REFERENCES

ActionAid. (2017). Malawi climate action report for 2016. ActionAid, Resilience and Economic Inclusion Team.

Adger, W. N., Huq, S., Brown, K., Conway, D., and Hulme, M. (2003). Adaptation to climate change in the developing world. *Progress in Development Studies*, 3, 179–195.

African Risk Capacity (ARC). (2016). Africa RiskView Monthly Bulletin | October 2016. Retrieved from: https://www.africanriskcapacity.org/wp-con tent/uploads/2016/12/ARV_ARVBulletinOct16_EN.pdf.

African Risk Capacity (ARC). (2017). African Risk Capacity – Response to ActionAid's Flawed Claims. Retrieved from: https://www.africanriskcapacity. org/wp-content/uploads/2017/07/African-Risk-Capacity-Action-Aid-Res ponse-Statement.pdf.

Ahmadalipour, A., and Moradkhani, H. (2017). Analyzing the uncertainty of ensemble-based gridded observations in land surface simulations and drought assessment. *Journal of Hydrology*, 555, 557–568.

Albris, K., Lauta, K. C., and Raju, E. (2020). Strengthening governance for disaster prevention: The enhancing risk management capabilities guidelines. *International Journal of Disaster Risk Reduction*, 47, 101647.

Alfieri, L., Salamon, P., Pappenberger, F., Wetterhall, F., and Thielen, J. (2012). Operational early warning systems for water-related hazards in Europe. *Environmental Science & Policy*, 21, 35–49.

Arndt, C., Schlosser, A., Strzepek., K., and Thurlow, J. (2014). Climate change and economic growth prospects for Malawi: An uncertainty approach. *Journal of African Economies*, 23(2), 83–107.

Awondo, S. N. (2019). Efficiency of region-wide catastrophic weather risk pools: Implications for African Risk Capacity insurance program. *Journal of Development Economics*, 136, 111–118.

Banda, M. (2016). More than half of Malawi's population need food relief – minister. Retrieved from: https://www.reuters.com/article/us-malawi-dro ught-idUSKCN0YG178.

Barnett, B. J., Barrett, C. B., and Skees, J. R. (2008). Poverty traps and index-based risk transfer products. *World Development*, 36(10), 1766–1785.

Becerra, O., Cavallo, E., and Noy, I. (2014). Foreign aid in the aftermath of large natural disasters. *Review of Development Economics*, 18(3), 445–460.

Benson, C., and Clay, E. (2004). *Understanding the economic and financial impacts of natural disasters*. Washington, DC: The World Bank.

Birkmann, J. (2011). First-and second-order adaptation to natural hazards and extreme events in the context of climate change. *Natural Hazards*, 58, 811–840.

Bouwer, L.M. (2019). Observed and projected impacts from extreme weather events: implications forloss and damage. In: *Loss and damage from climate change. Concepts, methods and policy options.* Springer.

Braman, L. M., van Aalst, M. K., Mason, S. J., Suarez, P., Ait-Chellouche, Y., and Tall, A. (2013). Climate forecasts in disaster management? Red Cross flood operations in West Africa, 2008. *Disasters*, 37(1), 144–164.

Brown, M. E., Funk, C. C., Galu, G., and Choularton, R. (2007). Earlier Famine Warning Possible Using Remote Sensing and Models. *Eos, Transactions American Geophysical Union*, 88(39), 381–382.

Calmanti, S., Syroka, J., Jones, C., Carfagna, F., Dell'Aquila, A., Hoefsloot, P., Kaffaf, S., and Nikulin, G. (2012). Model based climate information on drought risk in Africa. EGU General Assembly. Retrieved from: https://ui.adsabs.harvard.edu/abs/2012EGUGA..1410534C/.

Challinor, A.J., Koehler, A.K., Ramirez-Villegas, J., Whitfield, S., and Das, B. (2016). Current warming will reduce yields unless maize breeding and seed systems adapt immediately. *Nature Climate Change*, 6(10), 954–958.

Chamdimba, H. B. N., Mugagga, R. G., an Ako, E. O. (2020). Climate Change Mitigation and Adaptation through Anaerobic Digestion of Urban Waste in Malawi: A Review. *Journal of Energy Research and Reviews*, 4(4), 44–57.

Clarke, D., and Hill, R. (2013). Cost-benefit analysis of the African risk capacity facility. International Food Policy Research Institute (IFPRI) Discussion Paper 1292.

Clarke, D., and Wren-Lewis, L. (2016). Solving commitment problems in disaster risk finance. World Bank Policy Research Working Paper, (7720). Washington, DC: World Bank. Retrieved from: https://openknowledge.worldbank.org/handle/10986/24638.

Clarke, D. J., Mahul, O., Poulter, R., and Teh, T. L. (2017). Evaluating sovereign disaster risk finance strategies: a framework. *The Geneva Papers on Risk and Insurance-Issues and Practice*, 42(4), 565–584.

Conway, G. (2009). The science of climate change in Africa: Impacts and adaptation. Grantham Institute for Climate Change Discussion Paper, 1, 24.

Coughlan de Perez, E., van den Hurk, B. J. J. M., Van Aalst, M. K., Jongman, B., Klose, T., and Suarez, P. (2015). Forecast-based financing: an approach for catalyzing humanitarian action based on extreme weather and climate forecasts. *Natural Hazards and Earth System Sciences*, 15(4), 895–904.

Coughlan de Perez, E., van den Hurk, B., van Aalst, M. K., Amuron, I., Bamanya, D., Hauser, T., Jongma, B., Lopez, A., Mason, S., Mendler de Suarez, J., Pappenberger, F., Rueth, A., Stephens, E., Suarez, P., Wagemaker, J., and Zsoter, E. (2016). Action-based flood forecasting for triggering humanitarian action. *Hydrology and Earth System Sciences*, 20(9), 3549–3560.

Cummins, J.D., and Mahul, O. (2008). *Catastrophe risk financing in developing countries: principles for public intervention.* Washington, DC: The World Bank.

Dercon, S., and Christiaensen, L. (2011). Consumption risk, technology adoption and poverty traps: Evidence from Ethiopia. *Journal of Development Economics*, 96(2), 159–173.

Deressa, T. T., Hassan, R. M., Ringler, C., Alemu, T., and Yesuf, M. (2009). Determinants of farmers' choice of adaptation methods to climate change in the Nile Basin of Ethiopia. *Global Environmental Change*, 19(2), 248–255.

Easterling, W. E., Aggarwal, P. K., Batima, P., Brander, K. M., Erda, L., Howden, S. M., Kirilenko, A., Morton, J., Soussana, J. F., Schmidhuber, J. and Tubiello, F. N. (2007). *Food, fibre and forest products. Climate Change 2007: Impacts, Adaptation and Vulnerability.* Cambridge and New York: Cambridge University Press.

Ebi, K. L., Teisberg, T. J., Kalkstein, L. S., Robinson, L., and Weiher, R. F. (2004). Heat Watch/Warning Systems Save Lives: Estimated Costs and Benefits for Philadelphia 1995-98. *Bulletin of the American Meteorological Society*, 85(8), 1067–1074.

FAOSTAT (2017). Malawi Country Profile. Food and Agriculture Organization of the United Nations. Retrieved from: http://www.fao.org/faostat/en/#country/130.

Fouillet, A., Rey, G., Wagner, V., Laaidi, K., Empereur-Bissonnet, P., Le Tertre, A., Frayssinet, P., Bessemoulin, P., Laurent, F., De Crouy-Chanel, P., Jougla, E., and Hémon, D. (2008). Has the impact of heat waves on mortality changed in France since the European heat wave of summer 2003? A study of the 2006 heat wave. *International Journal of Epidemiology*, 37(2), 309–317.

Galindo, G., and Batta, R. (2013). Prepositioning of supplies in preparation for a hurricane under potential destruction of prepositioned supplies. *Socio-Economic Planning Sciences*, 47(1), 20–37.

Gero, A., Méheux, K., and Dominey-Howes, D. (2011). Integrating disaster risk reduction and climate change adaptation in the Pacific. *Climate and Development*, 3(4), 310–327.

Ghesquiere, F., and Mahul, O. (2007). *Sovereign natural disaster insurance for developing countries: A paradigm shift in catastrophe risk financing.* Washington, DC: The World Bank.

Ghosh, S., Vidyasagaran, V., and Sandeep, S. (2014). Smart cyclone alerts over the Indian subcontinent. *Atmospheric Science Letters*, 15(2), 157–158.

Global Facility for Disaster Reduction and Recovery (GFDRR). (2009). Post disaster needs assessment. October 2008 Tropical Storm and Floods, Hadramout and Al-Mahara, Republic of Yemen

Global Facility for Disaster Reduction and Recovery (GFDRR). (2013). Post Disaster Needs Assessment. Summary Reports. Retrieved from: https://www.GFDRR.org/node/118.

Golnaraghi, M., Surminski, S., and Schanz, K. U. (2016). An integrated approach to managing extreme events and climate risks: Towards a concerted public–private approach. Zurich: The Geneva Association. Retrieved from: https://www.genevaassociation.org/media/952146/20160908_ecoben20_final.pdf.

Gouramanis, C., and MoralesRamirez, C. A. (2020). Deep understanding of natural hazards based on the Sendai framework for disaster risk reduction in a higher education geography module in Singapore. *International Research in Geographical and Environmental Education*, 1–20.

Government of Malawi. (2015). Malawi 2015 floods post disaster needs assessments report. Malawi: Ministry of Disaster Management Affairs.

Government of Malawi. (2016). The national resilience plan. Breaking the cycle of food insecurity in Malawi. Malawi: Office of the Vice President.

Greatrex, H., Hansen, J.W., Garvin, S., Diro, R., Blakeley, S., Le Guen, M., Rao, K.N. and Osgood, D.E. (2015). Scaling up index insurance for smallholder farmers: Recent evidence and insights. Report 14 Copenhagen: CCAFS. Retrieved from: https://www.shareweb.ch/site/Climate-Change-and-Environment/Publications/CGIAR%20-%20Scaling%20up%20index%20insurance%20for%20smallholder%20farmers.pdf.

Gros, C., Bailey, M., Schwager, S., Hassan, A., Zingg, R., Uddin, M.M., Shahjahan, M., Islam, H., Lux, S., Jaime, C. and de Perez, E. C. (2019). Household-level effects of providing forecast-based cash in anticipation of extreme weather events: Quasi-experimental evidence from humanitarian interventions in the 2017 floods in Bangladesh. *International Journal of Disaster Risk Reduction*, 41, 101275.

Gurenko, E. N. (Ed.). (2004). *Catastrophe risk and reinsurance: a country risk management perspective*. World Bank Publications.

Handmer, J., and Dovers, S. (2013). *Handbook of disaster policies and institutions: improving emergency management and climate change adaptation*. London: Routledge.

Harriman, L. (2014). Cyclone Phailin in India: Early warning and timely actions saved lives. Thematic Focus: Environmental Governance, Disasters and Conflicts. *Environmental Development*, 9, 93–100.

Hart, N. C., Washington, R., and Stratton, R. A. (2018). Stronger local overturning in convective-permitting regional climate model improves simulation of the subtropical annual cycle. *Geophysical Research Letters*, 45(20), 11334–11342.

Haug, R., and Wold, B. K. G. (2017). Social protection or humanitarian assistance: contested input subsidies and climate adaptation in Malawi. *Institute of Development Studies (IDS) Bulletin*, 48(4), 93–110.

Hochrainer, S., an Pflug, G. (2009). Natural disaster risk bearing ability of governments: Consequences of kinked utility. *Journal of Natural Disaster Science*, 31(1), 11–21.

Hochrainer-Stigler, S., Sharma, R. B., and Mechler, R. (2012). Disaster microinsurance for pro-poor risk management: evidence from South Asia. *Journal of Integrated Disaster Risk Management*, 2(2), 1–19.

Houghton-Carr, H., and Fry, M. (2006). The decline of hydrological data collection for development of integrated water resource management tools in southern Africa. Climate variability and change—hydrological impacts (proceedings of the fifth FRIEND World Conference, Havana, Cuba, November 2006), IAHS Publ 308, 51–55.

Intergovernmental Panel on Climate Change (IPCC). (2007). *Climate change 2007: Impacts, adaptation and vulnerability.* Contribution of working group II to the fourth assessment report of the intergovernmental panel on climate change. Cambridge: Cambridge University Press.

Intergovernmental Panel on Climate Change (IPCC). (2012). *Managing the risks of extreme events and disasters to advance climate change adaptation.* A Special Report of Working Groups I and II of the Intergovernmental Panel on Climate Change. Cambridge: Cambridge University Press.

Internal Displacement Monitoring Centre (IDMC). (2020). Global report on internal displacement. Retrieved from: www.internal-displacement.org.

International Federation of Red Cross and Red Crescent Societies (IFRC). (2009). World Disaster Report 2009: Focus on early warning, early action. Retrieved from: http://www.ifrc.org/Global/WDR2009-full.pdf.

Ireland, P. (2010). Climate change adaptation and disaster risk reduction: Contested spaces and emerging opportunities in development theory and practice. *Climate and Development*, 2(4), 332–345.

Iyahen, E., and Syroka, J. (2018). Managing risks from climate change on the African continent: The African risk capacity (arc) as an innovative risk financing mechanism. In: *Resilience: The science of adaptation to climate change* (pp. 243–252). Elsevier.

Joshua, M.K., Ngongondo, C., Chipungu, F., Monjerezi, M., Liwenga, E., Majule, A.E., Stathers, T., and Lamboll, R. (2016). Climate change in semi-arid Malawi: Perceptions, adaptation strategies and water governance. *Jàmbá: Journal of Disaster Risk Studies*, 8(3), 1–10.

Juswanto, W., and Nugroho, S. A. (2017). Promoting Disaster Risk Financing in Asia and the Pacific Policy Brief. Tokyo: Asian Development Bank Institute. Retrieved from: https://www.adb.org/sites/default/files/publication/227516/adbi-pb2017-1.pdf.

Katongole, C. (2020). The role of disaster risk financing in building resilience of poor communities in the Karamoja region of Uganda: Evidence from an experimental study. International Journal of Disaster Risk Reduction, 45, 101458.

Kazianga, H., and Udry, C. (2006). Consumption smoothing? Livestock, insurance and drought in rural Burkina Faso. *Journal of Development Economics*, 79(2), 413–446.

Kehinde, B. (2014). Applicability of risk transfer tools to manage loss and damage from slow-onset climatic risks. *Procedia Economics and Finance*, 18, 710–717.

Kellet, J., and Caravani, A. (2013). Financing disaster risk reduction. A 20 years story of international aid. Global Facility for Disaster Reduction and Recover (GFDRR) and Overseas Development Institute (ODI). London, UK and Washington, USA

Kelman, I., Mercer, J., & Gaillard, J. C. (Eds.). (2017). Editorial introduction to this handbook: Why act on disaster risk reduction including climate change adaptation. In: *The Routledge handbook of disaster risk reduction including climate change adaptation*. Abingdon: Routledge.

Khamis, Khamis M. (2006). Climate change and smallholder farmers in Malawi: Understanding poor people's experiences in climate change adaptation; A Report by ActionAid. London, UK: ActionAid International.

Knowlton, K., Kulkarni, S. P., Azhar, G. S., Mavalankar, D., Jaiswal, A., Connolly, M., Nori-Sarma, A., Rajiva, A., Dutta, P., Deol, B., Sanchez, L., Khosla, R., Webster, P. J., Toma, V.E., Sheffield, P., and Hess, J. J. (2014). Development and implementation of South Asia's first heat-health action plan in Ahmedabad (Gujarat, India). *International Journal of Environmental Research and Public Health*, 11(4), 3473–3492.

Kotir, J. H. (2011). Climate change and variability in sub-Saharan Africa: a review of current and future trends and impacts on agriculture and food security. *Environment, Development and Sustainability*, 13(3), 587–605.

Kreft, S., Eckstein, D., and Melchior, I. (2017). Global Climate Risk Index 2017. Who suffers most from extreme weather events? Weather-related loss events in 2015 and 1996 to 2015. Briefing paper. Bonn: Germanwatch e.V. Retrieved from: https://germanwatch.org/sites/germanwatch.org/files/pub lication/16411.pdf.

Krishnamurthy, P. K., Choularton, R. J., and Kareiva, P. (2020). Dealing with uncertainty in famine predictions: How complex events affect food security early warning skill in the Greater Horn of Africa. *Global Food Security*, 26, 100374.

Krzysztofowicz, R. (2001). The case for probabilistic forecasting in hydrology. *Journal of Hydrology*, 249(1–4), 2–9.

Lewis, J. (1999). *Development in disaster-prone places: Studies of vulnerability*. London: Intermediate Technology Publications.

Linnerooth-Bayer, J., Mace, M.J., and Verheyen, R. (2003). Insurance-related actions and risk assessment in the context of the UNFCCC. Background paper for UNFCCC workshop on insurance-related actions and risk assessment. Bonn: United Nations Framework Convention on Climate Change (UNFCCC).

Linnerooth-Bayer, J., Mechler, R., and Pflug, G. (2005). Refocusing disaster aid. *Science*, 309, 1044–1046.

Linnenluecke, M. K., Griffiths, A., and Winn, M. (2012). Extreme weather events and the critical importance of anticipatory adaptation and organizational resilience in responding to impacts. *Business Strategy and the Environment*, 21(1), 17–32.

Linnerooth-Bayer, J., and Hochrainer-Stigler, S. (2015). Financial instruments for disaster risk management and climate change adaptation. *Climatic Change*, 133(1), 85–100.

Linnerooth-Bayer, J., Surminski, S., Bouwer, L. M., Noy, I., and Mechler, R. (2019). Insurance as a response to loss and damage? In: *Loss and damage from climate change*. Springer.

Lobell D.B., Burke M.B., Tebaldi C., Mastrandrea M.D., Falcon W.P. and Naylor R.L. (2008). Prioritizing climate change adaptation needs for food security in 2030. *Science*, 319, 607–610.

Lodree, E. J. (2011). Pre-storm emergency supplies inventory planning. *Journal of Humanitarian Logistics and Supply Chain Management*, 1, 50–77.

Lucas, B. (2015). Disaster risk financing and insurance in the Pacific. GSDRC Applied Knowledge Services.

Lung, F. (2013). The African risk capacity in the context of growing drought resilience in Sub-Saharan Africa. *Bologna Center Journal of International Affairs*, 16, 56–69.

Malawi Government. (2015a). Malawi vulnerability assessment report. Malawi Vulnerability Assessment Committee.

Malawi Government. (2015b). Malawi 2015 floods post disaster needs assessment report, 111 p.

Malawi Government. (2016). Malawi NAP stocktaking report. Retrieved from: http://adaptation-undp.org/sites/default/files/uploaded-images/mal awi_nap_stocktaking_report_final_2016.pdf.

Mall, R. K., Srivastava, R. K., Banerjee, T., Mishra, O. P., Bhatt, D., and Sonkar, G. (2019). Disaster risk reduction including climate change adaptation over south Asia: challenges and ways forward. *International Journal of Disaster Risk Science*, 10(1), 14–27.

Manyena, S. B. (2012). Disaster and Development Paradigms: Too Close for Comfort? *Development Policy Review*, 30(3), 327–345.

Martinez-Diaz, L., Sidner, L. and McClamrock, J. (2019). The future of disaster risk pooling for developing countries: where do we go from here? Washington, DC: World Resources Institute Working Paper. Retrieved from: https://www.wri.org/publication/disaster-risk-pooling.

McSweeney, C., New, M., Lizcano, G., and Lu, X. (2010). The UNDP climate change country profiles. *Bulletin of the American Meteorological Society*, 91(2), 157–166.

Mechler, R. (2005). Cost-benefit analysis of natural disaster risk management in developing and emerging countries. Working Paper. Deutsche Gesellschaft fuer Technische Zusammenarbeit (GTZ), Eschborn.

Mercer, J. (2010). Disaster risk reduction or climate change adaptation: are we reinventing the wheel? *Journal of International Development: The Journal of the Development Studies Association*, 22(2), 247–264.

Miller, S., and Keipi, K. (2005). *Strategies and financial instruments for disaster risk management in Latin America and the Caribbean*. Washington, DC: Inter-American Development Bank.

Mita, N. (2016). The development of disaster risk financing strategies in Japan. In: *Presentation at OECD–ADBI seminar on disaster risk financing in Asia* (Vol. 24). Tokyo: Asian Development Bank Institute.

Mittal, N., Vincent, K., Conway, D., Archer van Garderen, E., Pardoe, J., Todd, M., Washington, R., Siderius, C., and Mkwambisi, D. (2017). Future climate projections for Malawi. Future Climate for Africa Brief. Retrieved from: https://media.africaportal.org/documents/malawi_climatebrief_v6.pdf.

Moench, M. (2009). Adapting to climate change and the risks associated with natural hazards: Methods for moving from concepts to action. In: *The Earthscan reader on adaptation to climate change*. London: Earthscan.

Munich Re. (2015). Loss events worldwide 1980–2014. Geo Risks Research, NatCatSERVICE, Münchener Rückversicherungs-Gesellschaft: Munich, Germany.

Mutanga, C., Zulu, E., and De Souza, R.M. (2012). Population dynamics, climate change, and sustainable development in Africa. Washington, DC: Population Action International, African Institute for Development Policy. Retrieved from: https://www.africaportal.org/publications/population-dyn amics-climate-change-and-sustainable-development-in-africa/.

Mutegi, J., Kabambe, V., Zingore, S., Harawa, R., and Wairegi, L. (2015). The fertiliser recommendation issues in Malawi: Gaps, challenges, opportunities and guidelines. The Malawi Soil Health Consortium, Malawi, 56.

Nalau, J., Handmer, J., Dalesa, M., Foster, H., Edwards, J., Kauhiona, H., Yates, L. and Welegtabit, S. (2016). The practice of integrating adaptation and disaster risk reduction in the south-west Pacific. *Climate and Development*, 8(4), 365–375.

National Statistical Office. (2015a). Statistical yearbook. National Statistical Office, Zomba, Malawi: Government of Malawi.

National Statistics Office (NSO). (2015b). Welfare Monitoring Survey (WMS) 2014), Zomba, Malawi: National Statistical Office.

National Statistics Office. (2017). Integrated Household Survey 2016–2017. National Statistics Office, Government of Malawi.

National Statistical Office. (2019). 2018 Population and Housing Census: Main Report. National Statistical Office.

Niang, I., Ruppel, O. C., Abdrabo, M. A., Essel, A., Lennard, C., Padgham, J., et al. (2014). Africa in climate change 2014: Impacts, adaptation, and vulnerability. In: V. R. Barros, C. B. Field, D. J. Dokken, M. D. Mastrandrea, K. J. Mach, T. E. Bilir, M. Chatterjee, K. L. Ebi, Y. O. Estrada, R. C. Genova, B. Girma, E. S. Kissel, A. N. Levy, S. Maccracken, P. R. Mastrandrea, and L. L. White (Eds), *Part B: Regional aspects. Contribution of Working Group II to the Fifth assessment report of the Intergovernmental Panel on climate change* (pp. 1199–1265). UK; New York, NY, Cambridge: Cambridge University Press.

Nkomwa, E. C., Joshua, M. K., Ngongondo, C., Monjerezi, M., and Chipungu, F. (2014). Assessing indigenous knowledge systems and climate change adaptation strategies in agriculture: A case study of Chagaka Village, Chikhwawa, Southern Malawi. *Physics and Chemistry of the Earth*, 67–69, 164–172.

Pacific Catastrophe Risk Assessment and Financing Initiative (PCRAFI). (2013).

Parry, M. L., Canziani, O., Palutikof, J., Van der Linden, P., and Hanson, C. (Eds.). (2007). Climate change 2007: Impacts, adaptation and vulnerability. In: *Working group II contribution to the fourth assessment report of the IPCC* (Vol. 4). Cambridge University Press.

Priya, S., Young, W., Hopson, T., and Avasthi, A. (2017). Flood risk assessment and forecasting for the GangesBrahmaputra-Meghna River Basins. Washington, DC: World Bank. Retrieved from: https://elibrary.worldbank.org/doi/abs/10.1596/28574.

Red Cross Red Crescent Climate Centre. (2013). Health risk management in a changing climate. Retrieved from: https://www.climatecentre.org/downloads/files/Case%20studies/CC_HMR%20brochure_A4_6%20web.pdf.

Reeves, J. (2017). The wrong model for resilience: How G7-backed drought insurance failed Malawi, and what we must learn from it. ActionAid. Retrieved from: https://actionaid.org/sites/default/files/the_wrong_model_for_resilience_final_230517.pdf.

Rogers, D. P., and Tsirkunov, V. V. (2013). *Weather and climate resilience: Effective preparedness through national meteorological and hydrological services.* Washington, DC: The World Bank.

Sai, F., Cumiskey, L., Weerts, A., Bhattacharya, B., and Haque Khan, R. (2018). Towards impact-based flood forecasting and warning in Bangladesh: A case study at the local level in Sirajganj district. Natural Hazards and Earth System Sciences Discussions, 1–20.

Sandstrom, S., and Juhola, S. (2017). Continue to blame it on the rain? Conceptualization of drought and failure of food systems in the Greater Horn of Africa. *Environmental Hazards*, 16(1), 71–91.

Schipper, E. L. F. (2009). Meeting at the crossroads? Exploring the linkages between climate change adaptation and disaster risk reduction. *Climate and Development*, 1(1), 16–30.

Scott, Z., Simon, C., McConnell, J., and Villanueva, P. S. (2017). Independent evaluation of African Risk Capacity (ARC) Final Inception Report. Oxford Policy Management. Retrieved from: http://iati.dfid.gov.uk/iati_documents/27844297.pdf.

Shisanya, S., and Mafongoya, P. (2016). Adaptation to climate change and the impacts on household food security among rural farmers in uMzinyathi District of Kwazulu-Natal, South Africa. *Food Security*, 8(3), 597–608.

Stringer, L. C., Dyer, J. C., Reed, M. S., Dougill, A. J., Twyman, C., and Mkwambisi, D. (2009). Adaptations to climate change, drought and desertification: local insights to enhance policy in southern Africa. *Environmental Science & Policy*, 12, 748–765.

Suckall, N., Fraser, E., and Forster, P. (2017). Reduced migration under climate change: Evidence from Malawi using an aspirations and capabilities framework. *Climate and Development*, 9(4), 298–312.

Surminski, S., Bouwer L.M., and Linnerooth-Bayer, J. (2016). How insurance can support climate resilience. *Nature Climate Change*, 6 (4): 332–333.

Surminski, S., and Tanner, T. (Eds.). (2016). *Realising the 'Triple dividend of resilience'*. Berlin: Springer.

Surminski, S., Panda, A., and Lambert, P.J. (2019). Disaster insurance in developing Asia: an analysis of market-based schemes. Asian Development Bank Economics Working Paper Series. Retrieved from: https://www.adb.org/sites/default/files/publication/528026/ewp-590-disaster-insurance-developing-asia.pdf.

Swiss Re. (2012). Microinsurance-risk protection for 4 billion people. SIGMA No 6/2010, Zurich: Swiss Re. Retrieved from: https://www.findevgateway.org/sites/default/files/publications/files/mfg-en-paper-microinsurance-risk-protection-for-4-billion-people-2010.pdf.

Szlafsztein, C. F. (2020). Extreme natural events mitigation: An analysis of the National Disaster Funds in Latin America. *Frontiers in Climate*, 2, 17.

Tanner, T., Surminski, S., Wilkinson, E., Reid, R., Rentschler, J. and Rajput, S. (2015). The triple dividend of resilience: Realising development goals through the multiple benefits of disaster risk management. London, UK: Overseas Development Institute and Washington, DC: The World Bank. Retrieved from: http://www.odi.org/sites/odi.org.uk/files/odi-assets/publications-opinion-files/10103.pdf.

Tesfaye, K., Gbegbelegbe, S., Cairns, J. E., Shiferaw, B., Prasanna, B. M., Sonder, K., Boote, K., Makumbi, D. and Robertson, R. (2015). Maize systems under climate change in sub-Saharan Africa: Potential impacts on production and food security. *International Journal of Climate Change Strategies and Management*, 7(3), 247–271.

Thiboult, A., Anctil, F., and Ramos, M. H. (2017). How does the quantification of uncertainties affect the quality and value of flood early warning systems? *Journal of Hydrology*, 551, 365–373.

Thomalla, F., Downing, T., Spanger-Siegfried, E., Han, G., and Rockström, J. (2006). Reducing hazard vulnerability: towards a common approach between disaster risk reduction and climate adaptation. *Disasters*, 30(1), 39–48.

Trenberth, K. E., Jones, P. D., Ambenje, P., Bojariu, R., Easterling, D., Klein Tank, A., Parker, D., Rahimzadeh, F., Renwick, J.A., Rusticucci, M., Soden, B., and Zhai, P. (2007). *Observations: surface and atmospheric climate change. Climate change 2007: The physical science basis.* Contribution of Working Group I to the Fourth Assessment Report of the Intergovernmental Panel on Climate Change. Cambridge University Press, Cambridge, UK and New York, NY, USA.

Tsegai, D., and Kaushik, I. (2019). Drought risk insurance and sustainable land management: what are the options for integration? In: *Current directions in water scarcity research* (Vol. 2). Elsevier.

United Nations Development Programme (UNDP). (2009). A climate risk management approach to disaster reduction and adaptation to climate change. In: *The Earths can reader on adaptation to climate change.* London: Earthscan.

United Nations Development Programme (UNDP). (2016). Human development reports. Retrieved from: http://hdr.undp.org/en/composite/HDI.

UNICEF. (2019). 2018/19 National budget brief: Making government budgets work for children in Malawi, United Nations.

UNISDR and UNDP. (2012). Disaster risk reduction and climate change adaptation in the Pacific: An institutional and policy analysis. Suva, Fiji: UNISDR, UNDP, 76. Retrieved from: http://www.unisdr.org/files/26725_26725drra ndccainthepacificaninstitu.pdf.

UNISDR. (2016). Terminology. Retrieved from: https://www.unisdr.org/we/ inform/terminology.

United Nations General Assembly (UNGA). (2015). The Sendai framework for disaster risk reduction 2015–2030. Sendai, Japan: United Nations.

USAID. (2016). Malawi El Nino mitigation fact sheet. Retrieved from: https:// www.usaid.gov/malawi/fact-sheets/malawi-el-ni%C3%B1o-mitigation-factsheet.

Wang, B., Luo, X., Yang, Y.M., Sun, W., Cane, M.A., Cai, W., Yeh, S.W. and Liu, J. (2019). Historical change of El Niño properties sheds light on future changes of extreme El Niño. *Proceedings of the National Academy of Sciences,* 116(45), 22512–22517.

Warner, K., Ranger, N., Surminski, S., Arnold, M., Linnerooth-Bayer, J., Michel-Kerjan, E., Kovacs, P., and Herweijer, C. (2009). *Adaptation to climate change: Linking disaster risk reduction and insurance.* Geneva: United Nations International Strategy for Disaster Reduction. Retrieved from: http://www.asocam.org/sites/default/files/publicaciones/files/e22a13cd8f9c3b1f4cde901fe60a10e0.pdf.

Watts, N., Amann, M., Ayeb-Karlsson, S., Belesova, K., Bouley, T., Boykoff, M., Byass, P., Cai, W., Campbell-Lendrum, D., Chambers, J. and Cox, P.M, et al. (2018). The Lancet Countdown on health and climate change: From 25 years of inaction to a global transformation for public health. *The Lancet,* 391(10120), 581–630.

Webster, P.J., Jian, J., Hopson, T.M., Hoyos, C.D., Agudelo, P.A., Chang, H.R., Curry, J.A., Grossman, R.L., Palmer, T.N. and Subbiah, A.R. (2010). Extended-range probabilistic forecasts of Ganges and Brahmaputra floods in Bangladesh. *Bulletin of the American Meteorological Society,* 91(11), 1493–1514.

Webster, P. J. (2013). Improve weather forecasts for the developing world. *Nature,* 493(7430), 17–19.

Weingärtner, and Wilkinson, E. (2019). Anticipatory crisis financing and action: concepts, initiatives, and evidence. London: Centre for Disaster Protection. Retrieved from: https://static1.squarespace.com/static/5c9d3c35ab1a625 15124d7e9/t/5d15b8b33e88310001540fb7/1561704641415/Evidence_ review_Anticipatory_Crisis_Financing_Action.pdf.

Wiseman, W., and Hess, U. (2007). Reforming humanitarian finance in Ethiopia: A model for integrated risk financing. Working paper. Rome: United Nations World Food Programme (WFP).

Wisner, B., Blaikie, P., Blaikie, P. M., Cannon, T., and Davis, I. (2004). *At risk: Natural hazards, people's vulnerability and disasters.* London: Routledge.

Wisner, B., Gaillard, J. C., & Kelman, I. (Eds.). (2012). *Handbook of hazards and disaster risk reduction.* Abingdon: Routledge.

World Bank. (2007) The Caribbean catastrophe risk insurance initiative; results of preparation work on the design of a Caribbean catastrophe risk insurance facility. Washington, DC: The World Bank.

World Bank. (2013). Building resilience: Integrating climate and disaster risk into development. The World Bank group experience. Washington, DC: The World Bank.

World Bank. (2019). Malawi overview. Retrieved from: https://www.worldbank.org/en/country/malawi/overview.

World Food Programme (WFP). (2016). Food insecurity Worsens in Malawi, needs increase in face of El Ninō.

Yang, Y. E., Ray, P. A., Brown, C. M., Khalil, A. F., and Winston, H. Y. (2015). Estimation of flood damage functions for river basin planning: A case study in Bangladesh. *Natural Hazards*, 75(3), 2773–2791.

Zimmerman, F. J., and Carter, M. R. (2003). Asset smoothing, consumption smoothing and the reproduction of inequality under risk and subsistence constraints. *Journal of Development Economics*, 71(2), 233–260.

CHAPTER 8

Harnessing Digital Financial Solutions

Digital finance can be considered as the delivery of financial services that are linked to digital payment systems through mobile phones, personal computers, the Internet or cards (Ozili 2018). It has the potential to exponentially expand the delivery of financial services to large segments of society, such as the poor, through innovative technologies such as mobile banking solutions, e-money systems and digital payment infrastructure (Siddik and Kabiraj 2020). Manyika et al. (2016) define digital finance as financial services that are delivered through digital infrastructure such as through mobile phone and Internet networks, that overtime discourage the use of cash and traditional brick-and-mortar banking systems.

According to Gomber et al. (2017), digital finance includes a range of new financial products, financial businesses, financial software delivered by financial technology companies and innovative financial service providers. Although there is no standard definition of digital finance, according to Ozili (2018), there is some consensus that digital finance "encompasses all products, services, technology and/or infrastructure that enable individuals and companies to have access to payments, savings, and credit facilities via the internet (online) without the need to visit a bank branch or without dealing directly with the financial service provider." Digital finance therefore allows people unlimited access to finance from distant places, which enables them to manage their funds at times of need, as

© The Author(s), under exclusive license to Springer Nature Switzerland AG 2021
M. Ahmed, *Innovative Humanitarian Financing*,
Palgrave Studies in Impact Finance,
https://doi.org/10.1007/978-3-030-83209-4_8

well as reducing the possibility of them falling into poverty (Klapper et al. 2016). Moreover, according to Gabor and Brooks (2017), digital finance presents new methods of expanding the inclusion of groups deemed poor or disadvantaged into mainstream financial activities therefore adding new layers to financial inclusion.

Digital financial inclusion is the use of, and digital access to, formal financial services by excluded and underserved populations (Lyman and Lauer 2015). It determines the possibility of access to formal financial services for the general population based on "the implementation of basic principles of digital interaction between financial intermediaries and consumers, the use of innovative financial products, services, digital channels, as well as customer service and fund raising systems" (Naumenkova et al. 2019). According to Ozili (2018), digital financial inclusion starts with the assumption that segments of the population that are excluded or underserved have some kind of formal bank account and therefore require digital access in order to enable them to carry out basic financial transactions remotely.

There are several benefits of digital finance. From an economic perspective, Manyika et al. (2016) suggest that the wide-ranging use of digital finance could increase the annual gross domestic product (GDP) of all economies by 6% by 2025. Similarly, a study conducted by Ghosh (2016) found that digital financial services such as mobile telephony have significant impacts on economic growth. Other studies conducted have also found that digital finance has a positive effect on financial inclusion and economic growth (Siddik and Kabiraj 2020). Several actors in the humanitarian ecosystem are exploring the use of digital financial solutions to respond to humanitarian crises. These include donors, non-governmental organizations, humanitarian aid agencies, financial service providers, banks, microfinance institutions, mobile network operators and government agencies and regulators (Gurung and Perlman 2018).

DIGITAL FINANCIAL SERVICES

Digital financial services are financial services such as payments, remittances and credit, that are accessed and delivered through digital channels (Agur et al. 2020). They include a variety of instruments such as those that are offered by more established institutions such as banks for example, debit and credit cards, in addition to new solutions that are built on cloud computing, digital platforms, distributed ledger technologies, crypto-assets and peer-to-peer applications—these new solutions are commonly referred to as financial technology (fintech) (Agur et al. 2020).

Fintech is an emerging niche within the financial industry that refers to technology applied to financial activities (Schueffel 2016). According to Leong and Sung (2018), fintech can also be considered as "any innovative ideas that improve financial service processes by proposing technology solutions according to different business situations." There are six fintech business models: (1) insurance services, (2) crowdfunding, (3) payment, (4) lending, (5) wealth management and (6) capital markets (Lee and Shin 2018). Fintech has been transforming the financial services industry at an unprecedented rate (Frost et al. 2019) and there has been strong growth in digital innovation in fintech in particular, in the past decade or so (Phan et al. 2020). Moreover, developments in fintech have impacted financial planning, financial well-being and economic inequality (Frame et al. 2019) and has the potential to enhance financial capabilities (Panos and Wilson 2020).

There are a number of reasons as to why fintech has been gaining popularity. One reason for the emergence of fintech could be due to the need for financial services at more affordable costs which can then provide mobility and faster pace (Anikina et al. 2016). Fintech providers are able to provide faster financial services with a seamless process which makes it easier for low income individuals to manage their financial responsibilities on a day-to-day basis (Ozili 2018). Another reason could be due to the global financial crisis of 2008 where according to Haddad and Hornuf (2019), people lost confidence in the financial system and were looking for an alternative to give them more assurance. However, according to Ozili (2018), despite the emergence of digital innovation and the supposed impact it has on the financial industry "the effect of digital innovation and fintech growth on the financial system are less understood."

To provide financial services to crisis-affected populations, digital financial solutions are increasingly being used in humanitarian settings to provide cash-based assistance (Gurung and Perlman 2018; Whisson and May 2016). Digital financial solutions come in many forms including electronic vouchers through mobile devices or cards, debit cards, prepaid bank cards and mobile phone-based wallets (Bemo et al. 2017). These solutions have not only yielded benefits such as giving those in crisis-affected populations more choice, dignity, empowerment and resilience

(Harvey and Bailey 2015), but they also helped crisis-affected populations receive aid faster post-crises. Furthermore, they also offer several advantages over distributing physical cash such as faster distribution, heightened security, more accountability and transparency and a scalable transfer model (Bemo et al. 2017).

CASH-BASED ASSISTANCE

In humanitarian settings, cash-based assistance has been defined as "the provision of money to individuals or households, either as emergency relief intended to meet their basic needs for food and non-food items, or services, or to buy assets essential for the recovery of their livelihoods" (ECHO 2013). The nature of humanitarian assistance is changing as more humanitarian organizations are shifting from conventional commodities-based in-kind assistance such as temporary shelters, clothing, food, water and medical care—to using more cash-based assistance approaches (Smith et al. 2018).

In both developed and developing countries, conditional and unconditional cash transfers have become an increasingly common element of social protection policies (Fiszbein and Schady 2009; Arnold et al. 2011) and in humanitarian contexts, the use of cash transfer programming has increased significantly in recent years Furthermore, it is being recognized more as an integral component of humanitarian response that can make effective use of scarce resources, boost the dignity and choice for crisis-affected populations as well as address multiple sectoral outcomes at the same time (Arnold et al. 2011; Creti and Jaspars 2006; Gairdner et al. 2011; Venton et al. 2015). Several actors in the humanitarian ecosystem are in agreement that cash can be an effective tool in meeting the needs of those in fragile contexts as it can integrate humanitarian response within the local economy. Multilateral organizations, national governments, non-governmental organizations and civil society groups are using cash-based approaches to deliver humanitarian assistance across all sectors either on their own or in combination with in-kind provision of good or services (Gairdner et al. 2011).

Studies conducted have found that cash-based assistance can be more effective and efficient than in-kind assistance if designed and managed well. For example, in Rwanda, it was found that the benefit to the local community for each dollar of cash-based assistance given to refugees in the area was between USD$1.51 and USD$1.95, in comparison with

USD$1.20 worth of benefit for every dollar of food aid (Doocy and Tappis 2017; GSMA 2017a; Taylor et al. 2016). Furthermore, according to a study conducted by the International Rescue Committee (IRC) on cash and non-food item programs that served less than 1,000 households—it was found that non-food item programs cost more per dollar compared to cash programs of the same scale (IRC 2016).

In countries that are middle and higher-income, cash programs are often implemented electronically either through bank transfers or via prepaid debit cards; however, in low-income countries, cash transfer programs often require the physical distribution of cash in small quantities to remote rural areas due to limited financial infrastructure (Aker et al. 2016). As such, practitioners and stakeholders in humanitarian response such as donors, non-governmental organizations and humanitarian aid agencies are increasingly exploring the use of new technologies to provide humanitarian cash-based assistance (Gurung and Perlman 2018).

MOBILE MONEY

Of the digital financial solutions that have been used, mobile money is the most popular tool for digital cash transfers. Mobile money has become an alternative to traditional financial institutions as it allows for faster, cheaper, more convenient and easily accessible transactions, particularly for those living in rural and remote areas (Llanto et al. 2018). Furthermore, it eliminates transaction costs (i.e. transport, time, convenience) and enables more efficient and effective financial services (Zwendu 2014). Between 2014 and 2017, the portion of account owners sending or receiving payments rose from 67 to 76% globally and from 57 to 70% in the developing world (Demirguc-Kunt et al. 2018). Moreover, the largest growth in new and active mobile accounts has been in South Asia and sub-Saharan Africa (Alexander et al. 2017; Pasti 2019; Manyika et al. 2016; World Bank 2018).

Digital financial services via mobile devices have become an important tool to facilitate the financial inclusion of previously unbanked populations in developing countries (Kim et al. 2018) and an innovative means to provide financial services to excluded people (Sihvonen 2006). Mobile financial services encompass a variety of financial services such as mobile banking, mobile payment, mobile money transfer and mobile international remittance services (Kim et al. 2018). Mobile banking is a service providing customers with a channel to interact with a bank through a mobile device (Barnes and Corbitt 2003) and it refers to financial transactions that are conducted over a mobile device which has the potential to reach more people at a lower cost and is more convenient than traditional banking services that rely upon fixed brick and mortar branches (Haworth et al. 2016). Mobile payments, which is increasingly being used in developing countries, refer to the use of mobile devices to make payments for goods or services either at the point of sale or remotely (Siegel et al. 2011). In instances where users have reduced access to bank accounts but have a high demand for sending and receiving money between people, mobile money transfers are often used (Kim et al. 2018). Mobile international remittance services refer to international money transfers often used by migrant workers sending money to relatives in their home country whereby the remittances are received through a mobile device.

Advancements in mobile technology and the worldwide spread of mobile phones have caught the attention of humanitarian organizations and philanthropic institutions that are concerned with improving access to financial services for the poor globally (Duncombe and Boateng 2009). According to Maurer (2012), the interest around mobile payments can be explained by the convergence of the following three factors:

1. Increased interest from financial and communications service providers in improving fee-based revenue;
2. An awareness that information and communications technology is able to reach deeper into the Global South compared to many other institutions and industries due to the relatively low infrastructural requirements and light footprint, as compared to laying cables or building bank branches; and
3. More attention given to microfinance and the problems associated with the lack of access to financial services in general following the 2006 Nobel Peace Prize awarded to Grameen Bank founder, Mohammed Yunus.

Technology has enabled the delivery of financial services in ways that overcome traditional physical and economic barriers to financial inclusion (Gammage et al. 2017; Haider 2018; Mujeri and Azam 2018). This is particularly true when it comes to mobile phones and other digital access points, that have brought financial services directly to people. Some studies have shown that mobile-based financial services can increase pathways to financial inclusion in developing countries for low-income populations (Hinson 2011; Maurer 2012); moreover, these services have not only improved financial inclusion but have also helped overcome infrastructural constraints (Allen et al. 2014; Hinson 2011; Maurer 2012).

For example, in Kenya, millions of people are subscribed to a service called M-PESA that enables people to send money to friends and relatives cheaply and securely, through mobile phones. M-PESA is a mobile phone money transfer service that uses text messages and a network of retail agents as cash in and cash out points (Mas and Morawczynski 2009; Jack and Suri 2010). In the Philippines, people send money to family members who live on remote islands through a mobile transfer service called Globe GCASH (Maurer 2012). Also, mobile phones as a payment instrument have also been used in several fragile contexts to provide humanitarian cash-based assistance (Gurung and Perlman 2018).

Digital financial solutions such as mobile money also support international remittances, which as discussed in the previous chapter is an important source of income for those impacted by crises and in many cases is a vital lifeline. Prior to the COVID-19 pandemic, the size of cross-border remittances to developing countries had surpassed foreign direct investment (Pazarbasioglu et al. 2020). Furthermore, in refugee settings such as camps, both formal and informal remittance channels are increasingly being supported by new forms of mobile communication technologies that are making remittances easier to send. Given the growth of mobile banking and mobile remittances, these transnational financial flows are more likely to assume greater importance for refugee populations (Jacobsen 2012).

For example, in Uganda, the Bidi Bidi Refugee camp, which is mostly comprised of Sudanese refugees and has an increasingly growing refugee population, humanitarian aid agencies and mobile network operators have collaborated to provide humanitarian cash transfers by transferring funds to mobile network operator accounts from aid agencies bank accounts (Casswell and Frydrych 2017). The mobile network operator then transfers funds from its account to the mobile money account linked to the beneficiary's SIM card, where the beneficiaries are then able to cash-out from mobile money agents or spend the funds at a merchant—this is dependent on how much each aid agency decides to give the recipients (Casswell and Frydrych 2017).

DISTRIBUTED LEDGER TECHNOLOGIES AND BLOCKCHAIN

In humanitarian settings, there are several existing barriers that restrict the use of digital financial services. Often there are limiting policy requirements, poor network infrastructure and social barriers, that hinder the delivery of inclusive digital financial services in humanitarian response (Gurung and Perlman 2018). For instance, a lack of understanding may be a limiting factor for people to use mobile money, Ethiopia is a case in point where it was found that users who experienced difficulty with their mobile money PINs were 57% less likely to use their accounts again (Bailey 2017). To address these issues, new and innovative technologies such as blockchain technology are being explored and piloted by several humanitarian organizations.

Distributed ledger technology (DLT) and blockchain have been the subject of significant curiosity and criticism in the humanitarian sector. DLTs are essentially ledgers that are independently and identically distributed across multiple computing devices/entities known as nodes and are collections of accounts of information that are often financial (Coppi and Fast 2019). Each ledger is then copied continuously creating multiple and identical copies of the ledger.

Blockchain is a DLT that enables a range of complex digital interactions between entities without the verification and authentication practices that are usually conventionally provided by trusted third parties (Audia 2018). Blockchain technology records transactions then shares them through a network of multiple users and it is a chain where transactions are recorded across a peer-to-peer network to ensure that the value which is transferred is not duplicated—which is the definitive characteristic that differentiates blockchain from other existing peer-to-peer networks and cloud solutions (Audia 2018; Zwitter and Boisse-Despiaux 2018). Blockchain is best known for being the technology that underpins the digital currency Bitcoin (Nakamoto 2009) as well as laying the foundation for cryptocurrencies and smart contracts (Zwitter and Boisse-Despiaux 2018). Although there are overlaps between technology clusters, the terms DLT and blockchain are not interchangeable (Schulz and Feist 2020) as according to Natarajan et al. (2017), "not all distributed ledgers necessarily employ blockchain technology."

As decentralized databases, blockchains can be used to store non-monetary assets such as ownership rights, custodianship, contracts, goods, personal information as well as transactions (Zwitter and Boisse-Despiaux

2018). However, from a technical and capacity perspective, the implementation of DLTs is likely to be challenging in countries where digital infrastructure and electricity requirements are not met (Schulz and Feist 2020). Therefore, the effectiveness of DLTs depends on "the strength of a country's (digital) infrastructure — the Internet, distributed and cloud computing, electricity supply, and digitized data, all of which power the blockchain, as well as the technological literacy of its population" (Aggarwal and Floridi 2019). These factors coupled with existing "digital divides" may lead to instances where the implementation of DLTs could exacerbate existing inequalities within and among societies (Schulz and Feist 2020).

According to Gartner (2018a), DLTs are expected to reach their full potential over the next five to ten years furthermore, Gartner (2018b) predicts that "blockchain's business value-add will grow to slightly over USD\$360 billion by 2026, then surge to more than USD\$3.1 trillion by 2030." Moreover, some of the applications for DLTs that are considered to be the most promising include financial transactions, asset and supply chain management, energy markets, decentralized peer-to-peer networks for the exchange and storage of data, social service provision and digital identities (GIZ 2019; Zwitter and Herman 2018).

Blockchain and Humanitarian Aid

Proponents of blockchain believe that it is the next humanitarian aid disrupter. According to Reinsberg (2019), blockchain is considered favorable because it has the potential to enhance financial inclusion, improve production processes and improve governance and transparency in the humanitarian sector. Moreover, blockchain could better facilitate the sharing of data between various humanitarian organizations in an easier and cheaper way, while also protecting sensitive information about beneficiaries (Galen et al. 2018; GSMA 2017b). Given that donors have their own agendas, it is also argued that blockchain could help reduce bureaucracy and facilitate coordination between donors, which will in turn identify funding gaps in the humanitarian sector overall (Galen et al. 2018; Pisa and Juden 2017).

Blockchain technology has been used in various areas and for different purposes such as financial inclusion, remittances, improving donation transparency, reducing fraud, tracking the support to beneficiaries from multiple sources, micro-insurance, cash programming, grant management and cross-border transfers, among other uses (Coppi and Fast 2019). In terms of humanitarian uses, blockchain technology does not interact directly with the beneficiaries of humanitarian assistance, instead, it is used for the processes that take place on the backend to generate cost savings, increased transparency, traceability of information flows and application to financial transfers or supply chains (Ko and Verity 2016; Coppi and Fast 2019; Thylin and Novelo Duarte 2019). At the systemic level, blockchain powered tools could also be used to manage information, coordinate aid delivery, manage crowdfunding, track supply chains for cash transfer programming and help to enhance humanitarian financing overall (Riani 2018).

In humanitarian contexts enabling cash transfers is challenging due to the lack of diverse banking systems and the amount needed to cover service fees as conventionally, cash is usually transferred through wire transfers to bank accounts (Thylin and Novelo Duarte 2019) However, by conducting transfers on blockchain networks, both the sender and receiver are able to share a common infrastructure for real-time transactions therefore contributing to cash transfers that are more traceable, secure and lower-cost (Ko and Verity 2016). This is particularly important in times of crises as the ability to send and receive funds at a fast rate is crucial in critical situations, blockchain could therefore support the efforts of humanitarian actors to increase the use of cash transfers in humanitarian responses (Thylin and Novelo Duarte 2019).

Lack of access to financial systems and the lack of ability to transfer assets cause significant constraints on the economic activities of crisis-affected populations during peacetime and during crises (Thylin and Novelo Duarte 2019), blockchain technology is commonly suggested as having the ability to play an important role in the transfer of financial assets (Coppi and Fast 2019). Over the past few years, aid agencies and international organizations such as the United Nations High Commission for Refugees (UNHCR), United Nations Office for Project Services (UNOPS), United Nations Women and the World Food Programme (WFP), have been experimenting with the use of blockchain to distribute and monitor financial aid, to create digital identities, to secure land registries and to allow women access to microloans (Seyedsayamdost and Vanderwal 2020).

For example, in January 2017, the WFP piloted the "Building Blocks" project which used blockchain to deliver food and cash assistance to marginalized people in the Sindh province of Pakistan (UN Blockchain

2017), the project was also replicated in Jordan and Tunisia (Seyed-sayamdost and Vanderwal 2020). The intention of the project was to enhance collaboration across the humanitarian sector and use blockchain technology to make the WFP's cash-based transfers more efficient, transparent and secure (WFP 2018; Juskalian 2018). It was found that although blockchain technology reduces costs to the bulk nature of transactions and frees funds that can directly benefit end users, it does not eliminate intermediaries—states, international organizations or financial institutions (Seyedsayamdost and Vanderwal 2020). Instead, existing power structures are perpetuated as the technology does not create horizontal governance structures but rather empowers the organizations they embody for instance the WFP (Seyedsayamdost and Vanderwal 2020).

According to Makala and Anand (2018), blockchain is not a panacea for the ills of the humanitarian sector, it is a relatively new technology whose risks and potential are still largely being discovered and has not been thoroughly tested. However, going forward, the significant challenges and limitations of blockchain requires further study in order to better understand the implications blockchain technology will have on the humanitarian sector.

Conclusion

This chapter has briefly explored the opportunities associated with digital financial solutions in humanitarian settings. Although it is widely acknowledged that digital financial solutions in the long-term could enhance inclusion (Sahay et al. 2020), there are several risks inherent in their expansion, particularly in humanitarian contexts. For instance, a shift toward more digital financial services could worsen income and gender inequality, as women in developing countries are 23% less likely to access the Internet via mobile phones than men (Bahia and Suardi 2019).

One of the most popular digital financial solutions being used is digital cash transfers through mobile distribution. As the demand for mobile services continues to rise in humanitarian settings, its role in the provision of humanitarian aid could become more prominent. Further research should therefore be conducted on the expansion of digital cash transfers through mobile in the humanitarian sector and its implications to beneficiaries.

There are a number of pilot blockchain projects being implemented by humanitarian organizations to deliver aid and going forward, blockchain will no doubt play a larger role in the humanitarian sector. Blockchain is already being used to enhance gender equality, create secure digital identities, combat corruption, improve land tenure and property rights (Zwitter and Boisse-Despiaux 2018) as well as for other initiatives. However, blockchain is an emerging technology that is still in its infancy where its use and the implications of its use are still being understood; therefore, further research needs to be conducted to explore and evaluate the potential opportunities and challenges offered by blockchain to the humanitarian sector.

References

Aggarwal, N., and Floridi, L. (2019). The opportunities and challenges of blockchain in the fight against government corruption. 19th General Activity Report (2018) of the Council of Europe Group of States against Corruption (GRECO), Adopted by GRECO.

Agur, I., Peria, S. M., and Rochon, C. (2020). Digital financial services and the pandemic: Opportunities and risks for emerging and developing economies. *International Monetary Fund Special Series on COVID-19, Transactions*, 1, 2–1.

Aker, J. C., Boumnijel, R., McClelland, A., and Tierney, N. (2016). Payment mechanisms and antipoverty programs: Evidence from a mobile money cash transfer experiment in Niger. *Economic Development and Cultural Change*, 65(1), 1–37.

Alexander, A. J., Shi, L., and Solomon, B. (2017). How fintech is reaching the poor in Africa and Asia. EM Compass Note 34. Washington, DC.: International Finance Corporation (IFC), The World Bank Group. Retrieved from: https://openknowledge.worldbank.org/bitstream/handle/10986/30360/114396-BRI-EmCompass-Note-34-DFS-and-FinTech-Mar-28-PUBLIC.pdf.

Allen, F., Carletti, E., Cull, R., Qian, J., Senbet, L., and Valenzuela, P. (2014). The African financial development and financial inclusion gaps. *Journal of African Economies*, 23(5), 614–642.

Anikina, I. D., Gukova, V. A., Golodova, A .A., and Chekalkina, A. A. (2016). Methodological aspects of prioritization of financial tools for stimulation of innovative activities. *European Research Studies Journal*, 19(2), 100–112.

Arnold, C., Conway, T., and Greenslade, M. (2011). *Cash transfers: Evidence paper*. London: Department for International Development (DfID).

Audia, B. (2018). Implications of blockchain/DLT on the UN syste*m. In: The legal aspects of blockchain* (p. 6587). Copenhagen: United Nations Office for Project Services (UNOPS).

Bahia, K., and Suardi, S. (2019). *GSMA connected society: The state of mobile internet connectivity 2019*. London: GSMA. Retrieved from: https://www. gsma.com/mobilefordevelopment/wp-content/uploads/2019/07/GSMA-State-of-Mobile-Internet-Connectivity-Report-2019.pdf.

Bailey, S. (2017). *Electronic transfers in humanitarian assistance and uptake of financial services: A synthesis of ELAN case studies*. London: Overseas Development Institute. Retrieved from: https://www.odi.org/sites/odi.org.uk/files/resource-documents/11424.pdf.

Barnes, S. J., and Corbitt, B. (2003). Mobile banking: Concept and potential. *International Journal of Mobile Communications*, 1(3), 273–288.

Bemo, V. N., Aberra, D., Zimmerman, J., Lanzarone, A., and Lubinski, D. (2017). *Enabling digital financial services in humanitarian response*. Seattle, WA: Bill and Melinda Gates Foundation.

Casswell, J., and Frydrych, J. (2017). Humanitarian payment digitisation: Focus on Uganda's Bidi Bidi refugee settlement. United Kingdom: GSMA. Retrieved from: https://www.gsma.com/mobilefordevelopment/wp-content/uploads/2017/11/Humanitarian-Payment-Digitisation.pdf.

Coppi, G., and Fast, L. (2019). *Blockchain and distributed ledger technologies in the humanitarian sector*. London: Overseas Development Institute. Retrieved from: https://www.odi.org/sites/odi.org.uk/files/resource-documents/12605.pdf.

Creti, P., and Jaspars, S. (2006). *Cash transfer programming in emergencies*. Oxford: Oxfam.

Demirgüç-Kunt, A., Klapper, L., Singer, D., Ansar, S., and Hess, J. (2018). *The global findex database 2017: Measuring financial inclusion and the fintech revolution*. Washington, DC: The World Bank.

Doocy, S., and Tappis, H. (2017). Cash-based approaches in humanitarian emergencies: A systematic review. *Campbell Systematic Reviews*, 13(1), 1–200.

Duncombe, R., and Boateng, R. (2009). Mobile phones and financial services in developing countries: A review of concepts, methods, issues, evidence and future research directions. *Third World Quarterly*, 30(7), 1237–1258.

European Commission Directorate-General for Humanitarian Aid & Civil Protection (ECHO). (2013). The use of cash and vouchers in humanitarian crises: DG ECHO funding guidelines. Brussels: The European Commission. Retrieved from: https://ec.europa.eu/echo/files/policies/sectoral/ECHO_Cash_Vouchers_Guidelines.pdf.

Fiszbein, A., and Schady, N. (2009). Conditional cash transfers: Reducing present and future poverty. Washington, DC: World Bank, Policy Research report.

Frame, W. S., Wall, L., and White, L. J. (2019). Technological change and financial innovation in banking: Some implications for FinTech. In *Oxford handbook of banking*. Oxford: Oxford University Press.

Frost, J., Gambacorta, L., Huang, Y., Shin, H. S., and Zbinden, P. (2019). BigTech and the changing structure of financial intermediation. *Economic Policy*, 34(100), 761–799.

Gabor, D., and Brooks, S. (2017). The digital revolution in financial inclusion: International development in the fintech era. *New Political Economy*, 22(4), 423–436.

Gairdner, D., Mandelik, F., and Moberg, L. (2011). *We accept cash: Mapping study on the use of cash transfers in humanitarian recovery and transitional response*. Oslo: Norwegian Agency for Development Cooperation (NORAD).

Galen, D., Brand, N., Bourcherle, L., Davis, R., Do, N., El-Baz, B., Kimura, I., Wharton, K., and Lee, J. (2018). Blockchain for social impact: Moving beyond the hype. Stanford Graduate School of Business Center for Social Innovation and RippleWorks. Retrieved from: www.gsb.stanford.edu/sites/gsb/files/publication-pdf/study-blockchain-impact-moving-beyond-hype.pdf.

Gammage, S., Kes, A., Winograd, L., Sultana, N., Hiller, S., and Bourgault, S. (2017). *Gender and digital financial inclusion: What do we know and what do we need to know?* Washington, DC: International Center for Research on Women (ICRW).

Gartner. (2018a). The reality of blockchain. Retrieved from: https://www.gartner.com/smarterwithgartner/the-reality-of-blockchain/.

Gartner. (2018b). The CIO's guide to blockchain. Retrieved from: https://blogs.gartner.com/smarterwithgartner/the-cios-guide-to-blockchain/.

Ghosh, S. (2016). Does mobile telephony spur growth? Evidence from Indian states. *Telecommunications Policy*, 40(10–11), 1020–1031.

GIZ. (2019). *Blockchain potentials and limitations for selected climate policy instruments*. Bonn and Eschborn: GIZ.

Gomber, P., Koch, J. A., and Siering, M. (2017). Digital finance and FinTech: Current research and future research directions. *Journal of Business Economics*, 87(5), 537–580.

GSMA. (2017a). Landscape report: Mobile money, humanitarian cash transfers and displaced populations. Retrieved from: https://www.gsma.com/mobilefordevelopment/wp-content/uploads/2017/05/Mobile_Money_Humanitarian_Cash_Transfers.pdf.

GSMA. (2017b). *Blockchain for development: Emerging opportunities for mobile, identity and aid*. London: GSMA. Retrieved from: https://www.gsma.com/mobilefordevelopment/wp-content/uploads/2017/12/Blockchain-forDevelopment.pdf.

Gurung, N., and Perlman, L. (2018). Focus note: The role of digital financial services in humanitarian crises responses. Retrieved from: https://papers.ssrn.com/sol3/papers.cfm?abstract_id=3285931.

Haddad, C., and Hornuf, L. (2019). The emergence of the global fintech market: Economic and technological determinants. *Small Business Economics*, 53(1), 81–105.

Haider, H. (2018). *Innovative financial technologies to support livelihoods and economic outcomes. K4D helpdesk report*. Brighton: Institute of Development Studies (IDS).

Harvey, P., and Bailey, S. (2015). *Cash transfer programming and the humanitarian system: Background note for the high level panel on humanitarian cash transfers*. London, UK: Overseas Development Institute.

Haworth, A., Frandon-Martinez, C., Fayolle, V., and Simonet, C. (2016). Climate resilience and financial services. BRACED Working Paper. London: Overseas Development Institute. Retrieved from: https://www.odi.org/sites/odi.org.uk/files/odi-assets/publications-opinion-files/10316.pdf.

Hinson, R. E. (2011). Banking the poor: The role of mobiles. *Journal of Financial Services Marketing*, 15(4), 320–333.

IRC. (2016). Cost efficiency analysis: Non-food items vs. Cash transfers. Retrieved from: https://www.rescue.org/sites/default/files/document/958/nfidesignedbrieffinal.pdf.

Jack, W., and Suri, T. (2010). Mobile money: The economics of M-PESA. NBER Working Paper No. 16721. Retrieved from: http://www.nber.org/papers/w16721/.

Jacobsen, K. (2012). The economic security of refugees: Social capital, remittances, and humanitarian assistance. In *Global migration*. New York: Palgrave Macmillan.

Juskalian, R. (2018). Inside the Jordan refugee camp that runs on blockchain. MIT Technology Review 12.

Kim, M., Zoo, H., Lee, H., and Kang, J. (2018). Mobile financial services, financial inclusion, and development: A systematic review of academic literature. *Electronic Journal of Information Systems in Developing Countries*, 84(5), e12044.

Klapper, L., El-Zoghbi, M., and Hess, J. (2016). *Achieving the sustainable development goals: The role of financial inclusion*. Washington, DC: Consultative Group to Assist the Poor (CGAP).

Ko, V., and Verity, A. (2016). Blockchain for the humanitarian sector: Future opportunities, United Nations Office for the coordination of humanitarian affairs, digital humanitarian network. Retrieved from: https://reliefweb.int/report/world/blockchain-humanitarian-sector-future-opportunities.

Lee, I., and Shin, Y. J. (2018). Fintech: Ecosystem, business models, investment decisions, and challenges. *Business Horizons*, 61(1), 35–46.

Leong, K., and Sung, A. (2018). FinTech (Financial technology): What is it and how to use technologies to create business value in fintech way? *International Journal of Innovation, Management and Technology*, 9, 74–78.

Llanto, G. M., Rosellon, M. A. D., and Ortiz, M. K. P. (2018). E-finance in the Philippines: Status and prospects for digital financial inclusion (Discussion Paper Series No. 2018–22). Quezon City, Philippines: Research Information Department, Philippine Institute for Development Studies.

Lyman, T., and Lauer, K. (2015). What is digital financial inclusion and why does it matter? Consultative Group to Assist the Poor (CGAP).

Makala, B., and Anand, A. (2018). Blockchain and land administration. In *The legal aspects of blockchain* (pp. 131–150). Copenhagen: United Nations Office for Project Services (UNOPS).

Manyika, J., Lund, S., Singer, M., White, O., and Berry, C. (2016). Digital finance for all: Powering inclusive growth in emerging economies. San Francisco: McKinsey Global Institute. Retrieved from: http://www.mckinsey.com/global-themes/employment-and-growth/how-digital-finance-could-boostgrowth-in-emerging-economies.

Mas, I., and Morawczynski, O. (2009). Designing mobile money services: lessons from M-PESA. *Innovations: Technology, Governance, Globalization*, 4(2), 77–91.

Maurer, B. (2012). Mobile money: Communication, consumption and change in the payments space. *Journal of Development Studies*, 48(5), 589–604.

Mujeri, M. K., and Azam, S. E. (2018). Role of digital financial services in promoting inclusive growth in Bangladesh: Challenges and opportunities (Working Paper No. 55). Agargaon, Bangladesh: Institute for Inclusive Finance and Development. Retrieved from: http://inm.org.bd/wpcontent/uploads/2018/06/Working-Paper-55.pdf.

Nakamoto, S. (2009). Bitcoin: A peer-to-peer electronic cash system. Retrieved from: https://bitcoin.org/bitcoin.pdf.

Natarajan, H., Krause, S., and Gradstein, H. (2017). Distributed ledger technology (DLT) and blockchain. World Bank FinTech Note No. 1. Washington, DC: The World Bank.

Naumenkova, S., Mishchenko, S., and Dorofeiev, D. (2019). Digital financial inclusion: Evidence from Ukraine. *Investment Management and Financial Innovations*, 16(3), 194–205.

Ozili, P. K. (2018). Impact of digital finance on financial inclusion and stability. *Borsa Istanbul Review*, 18(4), 329–340.

Panos, G. A., and Wilson, J. O. (2020). Financial literacy and responsible finance in the FinTech era: Capabilities and challenges. *The European Journal of Finance*, 26(4–5), 297–301.

Pasti, F. (2019). State of the industry report on mobile money 2018. United Kingdom: GSMA. Retrieved from: https://www.gsma.com/r/wp-content/uploads/2019/05/GSMA-State-of-the-Industry-Report-on-MobileMoney-2018-1.pdf.

Pazarbasioglu, C., Mora, A. G., Uttamchandani, M., Natarajan, H., Feyen, E., and Saal, M. (2020). *Digital financial services*. Washington, DC: World Bank Group.

Phan, D. H. B., Narayan, P. K., Rahman, R. E., and Hutabarat, A. R. (2020). Do financial technology firms influence bank performance? *Pacific-Basin Finance Journal*, 62, 101210.

Pisa, M., and Juden, M. (2017). Blockchain and economic development: hype vs. reality. In Center for global development (CGD) policy paper. Washington, DC: Center for Global Development. Retrieved from: https://www.cgdev.org/publication/blockchain-and-economic-development-hype-vs-reality.

Reinsberg, B. (2019). Blockchain technology and the governance of foreign aid. *Journal of Institutional Economics*, 15(3), pp. 413–429.

Riani, T. (2018). Blockchain for social impact in aid and development. Humanitarian Advisory Group. Retrieved from: https://humanitarianadvisorygroup.org/blockchain-for-social-impact-in-aid-and-development/.

Sahay, M. R., von Allmen, M. U. E., Lahreche, M. A., Khera, P., Ogawa, M. S., Bazarbash, M., and Beaton, M. K. (2020). The promise of Fintech: Financial inclusion in the post COVID-19 Era. MCM Departmental Paper, International Monetary Fund. Retrieved from: https://www.imf.org/en/Publications/Departmental-Papers-Policy-Papers/Issues/2020/06/29/The-Promise-of-Fintech-Financial-Inclusion-in-the-Post-COVID-19-Era-48623.

Schueffel, P. (2016). Taming the beast: A scientific definition of fintech. *Journal of Innovation Management*, 4(4), 32–54.

Schulz, K., and Feist, M. (2020). Leveraging blockchain technology for innovative climate finance under the Green Climate Fund. Earth System Governance, 100084.

Seyedsayamdost, E., and Vanderwal, P. (2020). From good governance to governance for good: Blockchain for social impact. *Journal of International Development*, 32(6), 943–960.

Siddik, M. N. A., and Kabiraj, S. (2020). Digital finance for financial inclusion and inclusive growth. In *Digital transformation in business and society*. New York: Palgrave Macmillan.

Siegel, M., Schneidereit, F., and Houseman, D. (2011). Monetizing mobile: How banks are preserving their place in the payment value chain. Technical report, KPMG. Retrieved from: https://home.kpmg.com/ru/en/home/media/press-releases/2012/03/monetizing-mobile-how-banks-are-preserving-their-place-in-the-payment-value-chain.html.

Sihvonen, M. (2006). Ubiquitous financial services for developing countries. *The Electronic Journal of Information Systems in Developing Countries*, 28(1), 1–11.

Smith, G., McCormack, R., Jacobs, A., Chopra, A., Gupta, A. and Abell, T. (2018). The state of the world's cash report. Cash transfer programming in humanitarian aid. Oxford: Cash and Learning Partnership.

Taylor, J. E., Filipski, M. J., Alloush, M., Gupta, A., Valdes, R. I. R., and Gonzalez-Estrada, E. (2016). Economic impact of refugees. *Proceedings of the National Academy of Sciences*, 113(27), 7449–7453.

Thylin, T., and Duarte, M. F. N. (2019). Leveraging blockchain technology in humanitarian settings–opportunities and risks for women and girls. *Gender & Development*, 27(2), 317–336.

UN Blockchain. (2017). The world food programme wants to recruit ethereum in the fight against hunger. ETH News. Retrieved from: https://un-blockchain.org/2017/03/19/the-world-food-programme-wants-to-recruit-ethereum-in-the-fight-against-hunger/.

Venton, C. C., Bailey, S., and Pongracz, S. (2015). *Value for money of cash transfers in emergencies*. London: Department for International Development (DfID).

Whisson, I., and May, M. (2016). *Expecting the unexpected: why digital financial services can be a post disaster lifeline*. Washington, DC: Center for Financial Inclusion.

World Bank. (2018). Financial inclusion on the rise, but gaps remain, Global Findex database shows. Washington, DC.: The World Bank. Retrieved from: https://www.worldbank.org/en/news/pressrelease/2018/04/19/financial-inclusion-on-the-rise-but-gaps-remain-global-findex-database-shows.

World Food Programme. (2018). Blockchain for zero hunger: Building blocks. Retrieved from: https://innovation.wfp.org/project/building-blocks.

Zwendu, G. A. (2014). Financial inclusion, regulation and inclusive growth in Ethiopia. Working Paper 408. London: Overseas Development Institute.

Zwitter, A., and Boisse-Despiaux, M. (2018). Blockchain for humanitarian action and development aid. *Journal of International Humanitarian Action*, 3(1), 1–7.

Zwitter, A., and Herman, J. (2018). Blockchain for sustainable development goals: #Blockchain4SDGs— Report 2018. In Blockchain4SDGs workshop. University of Groningen, Leeuwarden.

CHAPTER 9

Conclusion

An effective humanitarian financing system should ensure that when a humanitarian crisis occurs—in order to save lives and prevent extreme suffering—those impacted receive immediate and timely relief. The global humanitarian financing system should therefore function as a safety net in times of crisis as well as, preparing for and preventing future crises while also taking into consideration humanitarian principles such as humanity, impartiality, neutrality and independence, to ensure innovative humanitarian financing models can uphold these principles and do not violate them.

Overall, the humanitarian financing system is built around several funding channels, however, governments are the greatest contributors to humanitarian aid on both a bilateral and multilateral level. Furthermore, in practice, funding for humanitarian aid is often channeled from government donors through several humanitarian organizations who then administer the aid on behalf of their donors and then distribute it to people in need of humanitarian assistance. One of the factors that is contributing to the lack of efficiency in the provision of humanitarian aid is donor fragmentation which is resulting in high transaction costs. This is due to donors and humanitarian organizations having their own programs that they implement, sectors they prioritize and procedures they follow.

© The Author(s), under exclusive license to Springer Nature
Switzerland AG 2021
M. Ahmed, *Innovative Humanitarian Financing*,
Palgrave Studies in Impact Finance,
https://doi.org/10.1007/978-3-030-83209-4_9

These inefficiencies tend to increase transaction costs which can therefore reduce the funding intended for vulnerable groups (de Vrij 2018). For example, according to Galen et al. (2018), aid organizations could potentially lose between 3% and up to 10% of funds due to transaction fees and inefficiencies.

Carpenter and Kent (2015) argue that one way of addressing the challenges of meeting growing humanitarian needs is to "recognise the added value and comparative advantages that those outside the so-called 'traditional' humanitarian sector can provide in strengthening efforts to manage risk and enhance preparedness, response and recovery." For instance, non-traditional donors such as those from the private sector contribute significantly to humanitarian financing. In the past few decades, we have not only seen the proliferation of private financing into the humanitarian sector at the national and international level but also, there has been an increase in innovative humanitarian funding models such as the ones discussed in this book as well as others. This is partly attributed to the explosion of humanitarian needs as globally, the number of humanitarian crises has increased for reasons such as environmental issues, forced displacement, protracted crises, among other reasons.

As mentioned in the introduction, the 2016 World Humanitarian Summit was a call to action to reform the global humanitarian sector and among other things, transform how humanitarian crises were being funded. One of the outcomes of the Summit was a recognition for the need to move beyond implementing short-term interventions and move toward contributing to more long-term sustainable solutions. Five years from the Summit and the global COVID-19 outbreak have also highlighted the need to reform the global humanitarian system and evaluate the role being played by international humanitarian organizations in responding to crises. Moreover, the COVID-19 pandemic is also creating immense humanitarian needs globally while also negatively impacting humanitarian aid supply chains. This coupled with the growing risks and uncertainty surrounding the pandemic is increasing needs and leaving populations in humanitarian contexts extremely vulnerable. Pre-COVID-19, there was great uncertainty about how the humanitarian funding gap could be filled, given the ongoing pandemic, there is even greater uncertainty and challenges ahead.

Also, since the Summit, the number of humanitarian crises has increased and significantly worsened, for example, the Yemen crisis is considered to be the world's worst humanitarian crisis by the United

Nations (UN). More than 3.6 million Yemenis have been displaced since the start of the crisis due to factors such as conflict, famine, multiple diseases and the lack of essential services (Hashim et al. 2021), according to the UN. Furthermore, most humanitarian crises have gone from bad to worse. Conflicts erupting in one country are spilling over into neighboring regions and becoming more complex and protracted. Countries experiencing severe crisis such as the Democratic Republic of Congo (DRC), Syria, Somalia, South Sudan and several countries in the Sahel are expected to be the top drivers of humanitarian aid (Jaff 2020), with most crises remaining significantly underfunded.

In recent times, there has been a proliferation of donors from the Global South as countries that were once aid recipients have become aid donors. Going forward, the rise of the Global South will not only reshape the world economy and shift power structures, but it will also impact the humanitarian funding system in many ways. Moreover, looking ahead, there will also likely be cuts to humanitarian aid budgets worldwide. The United Kingdom is a case in point as the U.K. government announced in November 2020 that it will move to reduce the humanitarian aid budget from 0.7 to 0.5% of national income leaving a financing gap of USD$6.2 billion in the official development assistance budget (Worley 2020). Where the U.K. has been a significant donor, this funding cut will no doubt have a significant impact on the humanitarian crises and programs it supports. Therefore, as more funding cuts are expected to global humanitarian aid budgets and the traditional humanitarian financing system is becoming increasingly stretched, the role of private finance will likely become more prominent.

Innovative humanitarian financing is an emergent field, however, there is little research that has been conducted in this space. This book is among the first to scrutinize innovative funding mechanisms being used in the humanitarian sector. The objective of this book was to evaluate these innovative financing instruments and mechanisms and assess whether they can reduce the humanitarian financing gap and adverse impacts of humanitarian crises. Overall, this book contributes to the scholarship on innovative humanitarian financing and provides insight into the evolving and increasingly complex humanitarian financing landscape.

The first part of the book introduced the book's aims, overall purpose and contemporary relevance. Through case studies, the second part of the book evaluated five innovative humanitarian financing mechanisms namely; advance market commitments, performance-based financing,

Islamic social finance, financial inclusion and disaster risk insurance. As these case studies have shown, the various financing mechanisms that were chosen for this book are differently suited to fund particular types of humanitarian crises. However, it is important to note that these funding models are in no way exhaustive as there were other innovative financing models that were not included as they were beyond the scope of this study. The final part of the book reviewed other trends in the humanitarian financing space such digital financial solutions and the role of blockchain in humanitarian aid.

This study was limited in part because of a shortage of data availability of innovative humanitarian financing mechanisms. Although the case study method proved valuable in being able to draw inferences, this method is limited in part because a case alone cannot provide a complete picture of the innovative humanitarian financing field. To extend this research further, more innovative humanitarian funding models should be analyzed, moreover, as the cases used in this research either focused on the healthcare or climate change sectors, further research should be conducted on humanitarian financing mechanisms used in other sectors. Research should also be conducted with industry practitioners to unearth insights and bridge the gap between academia and industry. Below is a set of recommendations that can be taken to enhance the ability of innovative humanitarian financing.

Innovative Humanitarian Financing: A Research and Policy Agenda

1. **Incorporating non-measurable metrics**
   Going forward, the contemporary humanitarian funding landscape will be marked by the growing role of the private sector and private sources of finance in order to fill the widening humanitarian financing gap. However, in order to appeal to donors, the results of humanitarian programs are often required to be easily quantifiable. The proliferation of private financing in the humanitarian sector will most likely bring a business-like approach whereby the need for measurable results will also be emphasized. One of the risks of innovative forms of humanitarian financing is the focus on such metrics where humanitarian organizations will be incentivized to prioritize projects and that are easily measurable. There is therefore a risk of more funding being allocated to projects and programs with clear

and measurable outcomes while other projects and programs that are not easily measurable being underfunded or side-lined.

However, governments, international humanitarian organizations and other global actors need to address and resolve the underlying issues and factors that are contributing to the growing number of humanitarian crises. The majority of the methods that can help reduce the impacts of crises and offset them—such as peace building and conflict resolution—are not easily measurable or quantifiable. Therefore, as more crises are becoming protracted and complex, the humanitarian funding models that are used should be tailored to match those needs and not just focus on clear measurable outcomes or programs where there is a high likelihood of success.

2. **Building resilience**

Overall, this book is part of a wider discussion about the importance and need for humanitarian aid to become more resilient to rapid shocks and fluctuations as a result of the changes that are occurring in the humanitarian sector.

Resilience is an increasingly popular construct shaping the humanitarian financing strategy in addition to disaster risk reduction. Resilience is a multi-layered concept that can broadly be defined as "the capacity that ensures adverse stressors and shocks do not have long-lasting adverse development consequences" (Constas et al. 2014a, b). According to the UNISDR Sendai Framework, countries and their partners should build resilience through response and reconstruction by "building back better" (Murphy et al. 2018). In essence, resilience requires agency to absorb, adapt and transform lives (Brück and d'Errico 2019). This emphasis on agency is significant for the study of innovative humanitarian financing for many reasons. For instance, the notion of resilience acknowledges how people who are being impacted by humanitarian crises are also agents in their own right therefore, rather than providing short-term aid and assistance, from a resilience perspective, supporting the aid beneficiaries' potential to deal more effectively with future crises and offsetting them is key.

More humanitarian organizations are designing and implementing programs that aim to enhance resilience and support those who are impacted by humanitarian crises such as climate-induced shocks. However, the measurement of resilience is a relatively novel and evolving area of research and practice (Bahadur et al. 2015;

Winderl 2014), and there is no uniform way to measure resilience. Nonetheless, a growing number of international humanitarian organizations have developed and prioritized resilience indicators as a key component of measuring the success of their programs (Alinovi et al. 2008; Schipper and Langston 2015). And with the increased attention on resilience, there has been a renewed interest in funding resilience, mitigation and adaptation projects by donors, multilateral institutions and organizations within the context of humanitarian response.

The majority of humanitarian aid policies focus on delivering immediate and short-term assistance where the focus is on relieving burden and providing relief. However, according to Alonso et al. (2012), the implementation of these policies in protracted environments has a tendency to create a dependency on aid. Innovative humanitarian financing mechanisms should therefore have more of a long-term strategy through enhancing resilience. For instance, innovative humanitarian financing mechanisms could help local communities build resilience against future crises by designing financing models that integrate resilience.

3. **Prioritizing localization**

Going forward, the humanitarian sector could use the COVID-19 pandemic as a catalyst to undertake much-needed reforms. Cuts anticipated to overseas development assistance and the ongoing pandemic offer an opportunity to transform the humanitarian financing landscape to undertake much-needed reforms such as the push for more "localization." The localization agenda, as first outlined in the 2016 World Humanitarian Summit, encourages national and international humanitarian organizations to facilitate more locally led responses and financing for humanitarian purposes (Murphy et al. 2018). For example, in Chapter six, the innovative funding model used was created in response to a local demand (from underserved populations) to access financial products and services that are aligned with local beliefs. Although there are issues with such innovative financing mechanisms, they are indicative of a need to shift the design and implementation of innovative humanitarian financing mechanisms to incorporate local traditions and values.

References

Alinovi, L., Mane, E., and Romano, D. (2008). Towards the measurement of household resilience to food insecurity: Applying a model to Palestinian household data. In: *Deriving food security information from national household budget surveys. Experiences, achievement, challenges* (pp. 137–152). Rome: FAO.

Alonso, J. A., Garcimartín, C., and Martin, V. (2012). 5. Aid, institutional quality, and taxation: Some challenges for the international cooperation system. In: *Development cooperation in times of crisis* (pp. 172–247). New York: Columbia University Press.

Bahadur, A., Wilkinson, E., and Tanner, T. M. (2015). Measuring resilience: An analytical review. Climate and Development.

Brück, T., and d'Errico, M. (2019). Reprint of: Food security and violent conflict: Introduction to the special issue. *World Development* 117, 167–171.

Carpenter, S., and Kent, R. (2015). The military, the private sector and traditional humanitarian actors. In: *The new humanitarians in international practice: Emerging actors and contested principles.* New York: Routledge.

Constas, M., Frankenberger, T., and Hoddinott, J. (2014a). Resilience measurement principles: Toward an agenda for measurement design. Food Security Information Network, Resilience Measurement Technical Working Group, Technical Series, 1.

Constas, M., Frankenberger, T., Hoddinott, J., Mock, N., Romano, D., Bene, C., and Maxwell, D. (2014b). A common analytical model for resilience measurement: causal framework and methodological options. Resilience Measurement Technical Working Group, FSiN Technical Series Paper, (2), 52.

de Vrij, A. (2018). Blockchain in Humanitarian Aid: A way out of poverty and famine? Master's thesis, Leiden University. Leiden: Netherlands. Retrieved from: https://openaccess.leidenuniv.nl/handle/1887/67036.

Galen, D., Brand, N., Boucherle, L., Davis, R., Do, N., El-Baz, B., Kimura, I., Wharton, K., and Lee, J. (2018). Blockchain for social impact: Moving beyond the hype. Stanford Graduate School of Business—Center for Social Innovation.

Hashim, H. T., Miranda, A. V., Babar, M. S., Essar, M. Y., Hussain, H., Ahmad, S., ... and Basalilah, A. F. M. (2021). Yemen's triple emergency: Food crisis amid a civil war and COVID-19 pandemic. Public Health in Practice, 2, 100082.

Jaff, D. (2020). Financing and resolving the ever-increasing humanitarian crises.*Medicine, Conflict and Survival*, 36(2), 129–131.

Murphy, R., Pelling, M., Adams, H., Di Vicenz, S., and Visman, E. (2018). Survivor-Led Response: Local recommendations to operationalise building back better. *International Journal of Disaster Risk Reduction*, 31, 135–142.

Schipper, E. L. F., and Langston, L. (2015). *A comparative overview of resilience measurement frameworks: Analysing indicators and approaches.* London: Overseas Development Institute.

Winderl, T. (2014). Disaster resilience measurements: Stocktaking of ongoing efforts in developing systems for measuring resilience. United Nations Development Programme (UNDP). Retrieved from: http://www.preventionweb.net/files/37916_disasterresiliencemeasurementsundpt.pdf.

Worley, W. (2020). Tony Blair: UK aid cuts and DFID closure a 'long-term strategic mistake'. Devex. Retrieved from: https://www.devex.com/news/tony-blair-uk-aid-cuts-and-dfid-closure-a-long-term-strategic-mistake-99415.

INDEX

© The Editor(s) (if applicable) and The Author(s), under exclusive
license to Springer Nature Switzerland AG 2021
M. Ahmed, *Innovative Humanitarian Financing*,
Palgrave Studies in Impact Finance,
https://doi.org/10.1007/978-3-030-83209-4

Lightning Source UK Ltd.
Milton Keynes UK
UKHW012236021121
393278UK00002B/5